"Drawing on solid credentials . . . Swan argues that understanding why people hunt involves understanding humankind's fundamental nature. Hunting is a spiritual ritual."
   —*Library Journal*

"*In Defense of Hunting* is the best and most thorough analysis of the morality of hunting there is, and is documented with invaluable reference material."
   —Walter E. Howard, professor emeritus of wildlife biology
      and vertebrate ecology, University of California at Davis

"Swan speaks with passion for the hunter, and with consummate reason to the anti-hunter. . . . If this reasoned defense of hunting hits its target, it will make us all better off."
   —Thomas McGuire, Ph.D., anthropologist,
      University of Arizona

"An exploration of our impulse to hunt and its value in society. The book calls upon hunters and nonhunters to work together to curb poachers, restore and preserve wildlife habitats, and ensure a future for both people and animals."
   —*Outdoors Unlimited*, Outdoor Writers of America

"A thorough response to those who denounce hunting."
   —*Kirkus Reviews*

"A philosophical but no-nonsense rebuttal to the anti-hunting movement."
   —Joe Doggett, *Houston Chronicle*

"A thoughtful look at why people hunt."
   —Dave Otto, Outdoors, *Green Bay Press-Gazette*

"Whether or not you're a hunter, you won't have to read very far before you realize this is an important contribution to this highly controversial subject. Swan is well qualified to write in depth on

this matter. . . . I was most impressed by the delicate balance he achieved between hunters and animal rights activists. . . . He offers sensible and logical solutions."
    —Dick Murdock, Outdoors, *Marin Independent Journal*

"A must for those who are interested in perpetuating the outdoor sport [of hunting]."
    —John Lohman, Outdoors, *The Forum*

"[Swan] writes passionately about the heart and soul of hunting . . . a meaningful act sparked by a spiritual love for nature."
    —Heritage Newspaper Group

In Defense of Hunting

James A. Swan

HarperSanFrancisco
An Imprint of HarperCollins*Publishers*

Other Books by
JAMES A. SWAN

Environmental Education (with William Stapp)
Building Networks (with Norman Gilroy)
Sacred Places
The Power of Place
Nature as Teacher and Healer
Bound to the Earth (with Roberta Swan)

IN DEFENSE OF HUNTING. Copyright © 1995 by James A. Swan.
All rights reserved. Printed in the United States of America. No part of this
book may be used or reproduced in any manner whatsoever without written
permission except in the case of brief quotations embodied in critical
articles and reviews. For information address HarperCollins
Publishers, 10 East 53rd Street, New York, NY 10022.

HarperCollins®, 📖®, and HarperSanFrancisco™ are trademarks of
HarperCollins Publishers Inc.

FIRST HARPERCOLLINS PAPERBACK EDITION PUBLISHED IN 1996
Book Design by Ralph Fowler
Set in Berling Roman

An Earlier Edition of This Book Was Cataloged As Follows:

Swan, James A.
    In defense of hunting / James A. Swan. —1st ed.
        p.   cm.
    Includes bibliographical references and index.
    ISBN 0-06-251029-0 (cloth)
    ISBN 0-06-251237-4 (pbk.)
        1. Hunting—Moral and ethical aspects.   2. Hunting—Psychological
    aspects.   I. Title.
SK14.3.S83   1994
799.2'01—dc20   94-27250
                                                                        CIP

        03 04 05 ❖ RRD 10

For my father,
who introduced me to
the magic of hunting

Contents

Acknowledgments

In writing a book, the final words that reach the eyes of the reader are sifted out from the energies, expertise, and knowledge of many, many people. I want to thank some of those who have been especially helpful on this project. Donald Michael provided many insights into the psychology of social movements and how to put statistical information into proper perspective. Jerry Longacre, Tom Brown, Stephen Kellert, Brian Hunter, Ted Nugent, Gary Kamia, Tyler Johnson, Robert Sudyam, Steve Johnson, Christine Thomas, Michael Billig, Jon Hooper, Kenny Cooper, and Medicine Grizzly Bear Lake made useful contributions in gathering research data, for which I am very appreciative. The staff at the Menninger Foundation was of great assistance in setting the record straight on what Karl Menninger really did have to say about hunting. Sara Pearl from the Harper Collins legal department made a number of useful contributions. My editor, Barbara Moulton, has been delightful to work with; she is remarkably skilled at working with writers to bring out their stories. And thanks, too, to Linda Allen, my literary agent, who is the kind of person I am proud to have represent me.

Introduction

The Hunter:
A Fallen Hero?

O ne of my earliest memories is of standing in my father's lap,
blowing the horn as we drove around in our black Ford
coupe, celebrating the end of World War II. I was two. Not long
afterward, my father went off for a few days to our cabin near
Grayling, in northern Michigan, and returned blowing the horn of
the car, this time with a large white-tailed buck tied to the roof.
Everyone in the neighborhood came out to see the deer and con-
gratulate Dad. A roast from that deer soon fed a group of friends,
and the skin was turned into a pair of buckskin gloves for my
mother, to match the buckskin jacket—made from skins of other
deer he had shot—that he had given her for a wedding present. To
me and to all our neighbors, my father was a hero, and I was very
proud.

Nearly fifty years later, as the wise old man in the hunter's
moon watched with patient, pale eyes, my teenage son and I
slipped away for an afternoon of roving through a field archery
course in the hills of southern Sonoma County, California. The
range straddled a creek, and the pungent aroma of decaying willow and alder leaves hung heavily in the cool fall air, the moist
stillness punctuated by the raucous calls of the scrub jays in the
live oaks and the proclamations of coveys of California quail scurry-
ing through blackberry and wild grape thickets.

Coming over the crest of a hill, we saw a fat black-tailed doe
feeding peacefully about ten yards to the side of a target forty

yards away. The doe abruptly picked up her head and eyed us, and we returned her gaze. A sense of excitement welled up in both deer and humans. My son and I had lethal weapons in hand. In that moment we entered into a continuum of human experience that has gone on for millions of years. The scent of the hunt was in the air, but on this day the blacktail was safe, as hunting is forbidden on the archery range, and the deer season for the region was long passed. My son and I moved to the mark, drew our compound bows, and shot. Two 25-12 aluminum-shaft arrows streaked toward a cardboard deer target pinned to three bales of straw. The arrows hit home in the optimum kill region, just behind the foreleg. Had the cardboard target been the doe, she would have gone down quickly, the heart and lungs mortally wounded, death from internal hemorrhaging coming within seconds. Still ten yards away, the doe looked up nonchalantly, as a somewhat bored spectator at an archery match. As we walked to the target, she took a couple of steps toward a blackberry thicket, then returned to browsing. The doe was now less than twenty yards away. I drew my bow as if to shoot at her. Her head came up. Our stares met across the shaft of my arrow. It was fall and the hunter's moon was full. Deer were fat for the coming winter. For at least 99 percent of human history, the correct thing to do under such circumstances would have been to kill the deer for food, skin, and perhaps hooves and dew claws for rattles. On that afternoon, however, politics intervened. I let down my bow and the doe dropped her head and went back to eating, waving her tail from side to side, almost spitefully, as if she knew she would not be shot. My son and I gathered our arrows and moved on to the next target. Overhead, a red-tailed hawk circled, watching for unwary field mice. Nearby, several black turkey vultures were riding the updrafts, their keen sense of smell, more than their eyes, searching for carrion. For them, hunting is still an instinctual pursuit unmediated by culture.

Some people hunt. Some do not. I am a hunter. When my father shot his deer in 1945, more than 25 percent of the men in the United States were hunters. In Michigan in those days I suspect the number was greater than 30 percent. Today, however, the hunter is an endangered species. In 1991, the National Survey of Fishing, Hunting, and Wildlife-associated Recreation reported that 14.1 million Americans sixteen years of age or older had hunted that year. This represents only 7.4 percent of the population over the age of fifteen, and a decline of over one-half million hunters from the 1985 survey that reported that 16.7 million Americans hunted. In California, where I now live, hunters are on an even more serious decline than the national average. In 1969, 750,000 hunting licenses were sold. In 1976, license sales dropped to 580,000, a 20 percent decline. Since then, despite a mushrooming population, there has been another 20 percent decline. As of 1993, only 415,000 other people—besides my son and me—bought hunting licenses in California. We hunters represent about 1 percent of the state's population, and our numbers are shrinking.

Who are hunters? The average hunter in the United States today is white, male, forty-two years old, lives in a small town or rural area in the middle of the United States (more than on either coast), has at least some college education, has a professional/managerial or services/labor job, and earns $43,120 per year. He hunts seven to ten days per year, primarily in his own state on private land within fifty miles of his home, with deer being by far the most popular animal hunted (10.3 million of the 14.1 million hunters say they hunt deer), and he spends about a thousand dollars a year on hunting licenses, equipment, travel, food, lodging, and so on. Compared with a decade ago, he is older, better educated, more likely to have a professional/managerial occupation, he earns more money, and he is more likely to hunt with a bow and arrow than ever before. As many as one-third of all hunters today use archery equipment, either solely or in addition to guns, the highest percentage since guns replaced the primitive spears, slings, and bows and arrows of our ancestors.[1]

Among those who stick with it, hunting can become almost an obsession, as evidenced by garages filled with boats and decoys, gun cabinets, engraved shotguns, hunting dogs, trophies on the walls, stacks of sporting magazines, wildlife paintings and carvings, stacked bales of hay in backyards for archery practice, and special clothing for each type of hunting—camouflage for waterfowl, wild turkey, and archery deer seasons, and bright orange for hunting big game with a rifle.

Even physical handicaps don't stop true hunters. Michigan bow hunter Rick Klein has only one arm, but that didn't stop him from bagging a nine-point white-tailed buck in 1992. He extends the bow outward with his right arm and anchors the arrow with his teeth. Bill Lanceford of Mississippi is a quadriplegic who hunts deer and wild turkeys from his wheelchair. His rifle is mounted in front of him, attached to the frame of his wheelchair. He wears a special football helmet with a face mask to keep him from being hit with the recoil when he fires, for the movement of his head is what aims the rifle, and his head is positioned directly behind the butt of the shortened stock. He activates the trigger by blowing on a special solenoid switch and gets his buck most years.

My father began teaching me to hunt when I was six or seven. Memories of walking with him through a carpet of scarlet, orange, and golden-brown leaves looking for fox and gray squirrels; slogging through waist-high marsh grasses in hopes of flushing out ring-necked pheasants; and sitting beside him in a duck blind of woven cattails and willows as the west wind swept across our backs and made the whitecaps dance on the gray November waves of Lake Erie; will forever be treasured in my mind.

As I trace back my family tree as far as I can on my father's side in Scotland, Norway, and Europe, our men have been strong hunters—ducks, geese, grouse, rabbits, deer, and reindeer were our quarry. The Swans, it seems, are part of the Gunn clan of Scotland, and our legacy is associated with the guarding of animals. Cattle were our main interest in Scotland, and in earlier times, in Scandinavia, we tended reindeer.

On my mother's side of the family, who came from Canada, there is both English and French ancestry, with equally long traditions of hunters among them. Another branch of my family tree includes mixed-blood Métis lineage, people who still trap, hunt, and fish in the forests of Ontario as they have done for thousands of years.

With this long chain of hunters watching me, as well as the hunter's moon on this October day, as my son and I trudged back to the car, I felt an ancient fire burning inside me, but its nostalgia was muted by sadness. Wearing leather shoes and driving a car with leather seats, along the way home we stopped for hamburgers and chicken, grown on factory farms perhaps five hundred or a thousand miles away. Heading south on Highway 101 we saw numerous glowing yellow deer eyes in the darkness beside the road and several carcasses of deer killed by speeding cars and trucks. At home, we watched television for a while and read the evening newspaper, acknowledging the rising tide of violence in the world. In our community, there are only a handful of hunters, as far as I know, but the majority of homeowners find their night's sleep more restful with a firearm hidden within a few feet of the bed. Nowadays it seems that people hunt one another more as hunting for animals wanes.

Nature is not what it was a century ago. There is good reason for concern about protecting endangered species of plants and animals: Modern civilization is gobbling them up at an alarming rate. Due to market hunting for profit and habitat destruction, in the nineteenth century many species of wildlife were decimated around the world, and some became extinct. Passenger pigeons once darkened the skies in teeming flocks of billions of birds that stretched from horizon to horizon. The last survivor, named Martha, died at the ripe old pigeon age of twenty-nine in the Cincinnati Zoo, on August 1, 1914. One month later, in that same zoo, the last Carolina parakeet also died. The dodo, the thirteen-foot-tall flightless moa bird of New Zealand, the giant elephant birds of Madagascar that laid eggs three feet in circumference, the Labrador duck of the North Atlantic, and the heath hen of New England disappeared due to market hunting and habitat loss. The American bison, once the most common ungulate on earth at 60 million or more, slipped to 500 in 1900 before it was saved from extinction. The California State animal is the grizzly bear, but like the Michigan wolverine, none remain in the wild. As rain forests are cut down, oceans are polluted, old-growth forests are clear-cut, and wetlands are swallowed up by roads, parking lots, and malls, many plant and animal species are going extinct. But not all nature is vanishing.

Hunting is a blood sport, but modern sport hunting has not caused the extinction of any animal species. In fact, since the early 1900s, many of the most popular game species in the United States have become more abundant, thanks to the curtailing of market hunting, the development of scientific wildlife management to determine seasons and limits, and help from revenues generated by the sales of hunting licenses and ammunition that have been used for habitat protection and restoration. In the 1920s, there were roughly 300,000 white-tailed deer in the United States. Today, there are more than 27 million. During the Great Depression, elk numbered 50,000; today there are at least twenty times that many. The pronghorn antelope population of the Dust Bowl days has grown from 25,000 to close to a million. The wild turkey population has skyrocketed from 30,000 in 1920 to more than 4 million. The United States bison herd, which in 1900 was no more than 500, with only 39 in the wild in Yellowstone National Park, now numbers 120,000 and is growing rapidly on farms, hunting preserves, state and federal parks, and wildlife sanctuaries. There were slightly more than a million Canada geese in the 1940s when my father started taking me out to the duck blind. Today, the Canada goose population is 2.5 million and growing. Wood ducks, which nest in trees, were nearly extinct in the late 1800s and early 1900s, due to clear-cut lumbering in the eastern forests, which decimated their nesting habitat. Wood ducks today are the most common breeding waterfowl in the United States, thanks to the massive campaign to put out human-made nesting boxes and to the refinement of modern forestry practices.[2]

The animals that are staging a comeback, like white-tailed deer, Canada geese, and mallards, are the most adaptable. Those who have specialized niches, like the notorious spotted owl of the Pacific Northwest, which prefers to nest only in old-growth evergreens, or the Florida cougar or panther, which resides in swamps (and more than 9 million acres of Florida swamps have been drained in the last two hundred years) have the lowest odds for surviving into the twenty-first century.

There is considerable news in the media about the loss of tropical rain forests worldwide, but not enough attention is given to the loss of one of the most valuable ecosystems of North America—wetlands. Bogs, swamps, and marshes are being gobbled up by farms, roads, parking lots, subdivisions, and malls at the alarming rate of 650 acres a day—over 300,000 acres a year. When the earliest European settlers arrived in North America there were approximately 215 million acres of wetlands in the lower forty-eight states. Nationwide, 53 percent of these original wetlands have disappeared. In California, Connecticut, and Illinois, 90 percent of the states' original wetlands are gone. It is little wonder then that in the last twenty years the duck population of North America has plummeted from 100 million to 62 million. In 1954, more than a million ducks wintered in Indiana. Forty years later, a winter census reported barely 9,000. A prime factor for the dramatic decline of ducks in Indiana is that the state has lost 87 percent of its wetlands, which ducks use for food, shelter and nesting. Dabbling ducks, which feed in shallow waters, have been especially hard hit by the drying up of North America. In the last twenty years, the ancestors of the black ducks that my father and I hunted on Lake Erie are 60 percent fewer in number. Nationwide, pintails are down 65 percent, and mallards are down more than 40 percent, despite widespread release of pen-raised mallards.

About 200 million animals are killed every year by hunters in the United States. The breakdown for some of the most common species is:

50 million doves
25 million rabbits and squirrels
25 million quail
20 million pheasant
10 million ducks
4 million deer
2 million geese
150,000 elk
21,000 black bears

To the nonhunter, these figures may seem high, but none of these animals is in danger of extinction due to hunting. And as any hunter will tell you, hunting today is heavily monitored, with many restrictions on species, seasons, weapons, methods, and locations. In California today, in order not to violate any game laws, I must wade through a 32-page booklet governing bird hunting, a 24-page pamphlet on state and federal areas, and a 20-page booklet for big game. In Vermont, hunting and fishing laws are condensed into an 80-page booklet the size of a dollar bill. Michigan's hunting and trapping guide is 30 pages, including ads, and special regulations for migratory birds are printed on a two-sided fold-out poster. Prior even to buying a hunting license, new hunters in most states now must take a special hunter education course on firearms safety, wildlife conservation, and wildlife identification. To hunt today, you cannot just get a gun and buy a license and shoot whatever you want, as was the case a century ago.

Despite the careful regulation of hunting and hunters, and the work of state and federal agencies and hunting organizations such as Ducks Unlimited, the Rocky Mountain Elk Foundation, Safari Club International, and Quail Unlimited to preserve habitats, today there is an antihunting sentiment that is stronger than ever before.

Hunting is no longer considered an unquestioned act of human nature or good citizenship. Whereas forty-five years ago the hunter was seen as a hero, today he or she is increasingly portrayed as a villain. Studies by Dr. Stephen Kellert at Yale University find that while 80 percent of Americans approve of hunting for putting meat on the table, 60 percent do not approve of hunting for recreation or sport, and 33 percent favor banning hunting altogether.

A new "subspecies" of human, the *animals rights activists*, has suddenly appeared, and they have a large following. Espousing the philosophies of Albert Schweitzer and Mahatma Gandhi, *Homo sapiens animal rightists* deplore hunting as savage brutality and cruelty, and some of them have no qualms about turning their feelings into action. In recent years nearly all states have had to pass special laws to protect hunters in the field from harassment by people with bullhorns, smoke bombs, and noise makers. The protests have even erupted in the tiny town of Guerneville, California, a two-block-long hamlet tucked among the giant redwoods that line the banks of the Russian River. Here, each spring, to brush away the cobwebs of the foggy, rainy winters of the coastal forest, the citizens of Guerneville hold a unique event called the Slugfest. The Slugfest is a fun-filled festival honoring the giant yellow, green, and brown banana slugs that can reach seven inches long in these parts. Slug races, contests for the largest slug, and a special cook-off of dishes made with this shelless escargot are among the festivities. Slug pizza, incidentally, is a perennial winner.

In recent years, the Slugfest has sometimes come closer to another interpretation of the name of the event—fisticuffs—as animal rights protesters have squared off against the town fathers to protest the treatment of slugs in the Slugfest. Carrying signs that read SLUGS HAVE RIGHTS TOO and SLUGS ARE SOME OF MY BEST FRIENDS, and making accusations of cruelty and sadism, the protestors march and shout just a few feet from kids showing off their pet slugs, crowds cheering for the fastest slugs, and local celebrities sampling slug cuisine. According to the organizers, all slugs participating in the event are released, unharmed, after the event, except those that are cooked, of course. As any California gardener knows, slugs are not an endangered species in Northern California! In fact, not that long ago, a diet of slugs saved many

Miwok, Pomo, and Yurok Indians, native to the coastal forest, from starvation during hard winters. During the Slugfest, fewer than a hundred slugs are killed for food, which is far fewer than the number that are squashed by the tires of the protesters' cars as they drive to and from the protest, but nonetheless, they have been showing up each year to express their constitutional rights of free speech in protesting the Slugfest.

For many people throughout history, the most seductive voice of Mother Nature at special times of the year has been the invitation to join in the quest to hunt and kill birds, mammals, reptiles, and fish. But times have changed. Opposition to hunting today is at an all-time high and shows no signs of decreasing. In fact, as the years pass, the goals of the antihunters increase. Some speak openly of the time in the not too distant future when they see all hunting banned forever. Hunters, they insist, are no longer heroes but sadists who enjoy the heinous torturing of helpless animals. As antihunting sentiments escalate, the tactics of obstruction become more desperate. In 1992, a bear hunter in British Columbia had a shot fired over his head by an animal rights protester. In 1993, in Kalamazoo, Michigan, a man was arrested on suspicion of killing eight hunters over a two-year period, coldly gunning them down in the woods. Just a few days later a Michigan archer was found dead in a thicket. The body had two bullet holes, one in the heart and one in the head. Although there is not yet any evidence, there are those who speculate that these murders may be the work of animal rights activists. Taxidermists receive threatening phone calls, and some shops have been burned down, moving some craftsmen to use unlisted phone numbers and refrain from public advertising. Companies that tan animal hides now have tight security systems with alarms and remote cameras to watch for intruders who might try to set fire to the hides or spray them with paint. Among the leaders of the antihunting movement, Christine Stevens, of the Animal Welfare Institute, has been quoted as saying, "We're very much against acts of violence, and we don't do anything illegal. But we don't condemn people who do either, so long as they don't do anything dangerous."

Today, living in Marin County, California, a few miles north of the Golden Gate Bridge, with San Francisco Bay on one side and the Golden Gate National Recreation Area, Muir Woods National Monument, and Mount Tamalpais State Park on the other, wildlife is all around us. Relatives of the Sonoma doe roam the hills, we know, as we wake to find that all the flowers in the planters on our patio have disappeared. The rough, uneven marks on the stems show that they were pulled off, not sharply cut off—unmistakable signs of a browsing animal without sharp incisors. Purple petunias seem to be a local favorite. California's destruction of forests and wetlands is disgraceful, and yet I can watch nearly as many species of wildlife as my ancestors did. The fat four-point buck that savors my wife's purple petunias has no fear of arrows in this neighborhood, where hunting is illegal. But recently he and a fawn were taken by a modern predator, a BMW, like the hundreds of other deer that are killed by cars each year in Marin County. There is no public land in Marin where deer hunting is legal, and as a result deer are everywhere. There are so many of them that firearms are not necessary for poaching here, because the deer are so numerous and have little fear of humans. A game warden recently caught a group of deer poachers in Marin who had been using high-speed cars to kill deer. In the trunk of a Cadillac with a special, reinforced front bumper, pulled over for speeding on country roads in the early morning hours, were three freshly killed deer.

Out for a jog beside San Francisco Bay, I pass flocks of mallards and pintails gathered in open pockets of the tidal marsh. I watch a man unleash his dog for a morning run. The Gordon setter, a breed cultivated for hunting, was excited at getting loose and tore off at breakneck speed through the salt grass. Soon the dog spied a killdeer perched near the shore. The dog rushed after the bird, which took off, loudly calling its name *killdeer* as it rose.

With the dog's owner screaming "Come! Come!" and running after his pet, the setter ran at full speed along the bank, leaping wildly into the air to try to catch the killdeer, already fifty feet above its head. A hundred yards farther, the bird disappeared into the fog overhead and the dog returned to its irate owner, wagging its tail madly but with its head hanging down, indicating that it realized it disobeyed its master. There was, however, a definite grin on the dog's face. It had enjoyed its disobedience thoroughly.

Since the beginning of time, men and women have hunted. Their tools and methods have varied considerably, but when the first breath of fall air comes rushing down from the north, and ghostly strings of geese pass across the face of the hunter's moon, a special kind of adrenaline begins to pump up from the deeper recesses of the instinctual psyche, calling up sentiments that link even modern people back to the Paleolithic and beyond. These quiet voices of nature's kinship are always there, but since we are the "thinking animal," we exercise more choice and control over our instincts than other animals do. Like the dog, today we "civilized" people live in controlled situations, with our little bit of precious free time, taking vacations in national parks and untamed lands in the hope of restoring our health and sanity. Hunting is not necessary for our survival, and because fewer and fewer people today hunt, in an age when there is so much violence and concern about ecology and animals becoming extinct, the hunter becomes an easy scapegoat for the troubles of society, whether he deserves the rap or not. Reviewing the studies of hunting trends over the last few decades, Professors Daniel Decker, Jody Enck, and Tommy Brown of Cornell University recently concluded: "Certainly without, and perhaps even with, extraordinary intervention efforts on a scale we've never seen before, hunting is going to continue to decline over the foreseeable future." [3]

Farther down the jogging path beside the Bay, I came upon a cat playing with a mouse it had just caught. Despite centuries of domestication, cats have retained more of their hunting skills than dogs have. We tolerate this carnality in Tabby because his propensity for mousing helps guard our health and property. We like to think of ourselves as having risen above the animal world, but that is an illusion, according to that wisest explorer of the unconscious mind, Carl Jung. From his years of studying the wilderness within, which is revealed in our dreams, Jung concluded that the healthy, primal symbols of the unconscious are the symbols of nature. The self is a stone or a gemstone. A tree represents the union of conscious and unconscious, its type giving us hints about our character. Symbols of animals in our dreams represent instincts in both their positive and negative aspects; the species telling us things about the physical nature of our instinctual selves and how we prefer to express it.

In modern society, there is a tendency to try to pave over the inner world as much as the outer world, and to deny our unconscious, as Freud revealed to us in his classic work *Civilization and Its Discontents*. The relationship between denial of unconscious energies and the symbols and illness has now been well documented. Within each of us, "a part of the soul is a leopard," Jung said. Fully appreciating this statement should cause those who deny they have a violent side to shiver. If in each of us there is a leopard, then our nature includes a deadly hunter, regardless if we are vegetarians or enjoy rare steaks. Faced with starvation, few people would opt to perish rather than kill an animal for food.

Just like my ancestors who danced around trees, made sacrifices to the spirits of the wild, and prayed to the spirit keepers of the animals before going on a hunt, my stomach gets empty and I must eat. Today, however, we stalk the supermarket for bargains. I can almost hear the deer chuckle as I walk to Safeway to gather vegetables grown to maturity in the Central Valley of Mexico, the water for irrigation being diverted from wetlands that support nesting ducks and spawning fish. Venison roast will never come from my neighborhood, at least legally. Instead, I will buy pen-raised pork, chicken, and beef, which have been transported often a thousand miles or more. Unless I am willing to pay top dollar, the meat will contain traces of chemical fertilizers, pesticides, and far more cholesterol than can be found in the deer who bed within fifty feet of my house after finishing off the petunias. Although I live beside San Francisco Bay, I worry, with good reason, about eating fish or shellfish that come from it. The California fishing-laws booklet advises that children and pregnant women, or women who hope to become pregnant, should not eat seafood from San Francisco Bay due to the high levels of mercury. So the shelves in the local supermarket display New Zealand orange roughy, lobster from Maine, shrimp from the Caribbean, mahi mahi from Hawaii, crabs from the Bering Sea, and recently in drought years, Norwegian salmon, rather than seafood harvested from local waters. When we replaced hunting with domestication of animals we got convenience and assurances of supply, which enabled more and more people to leave their food-producing jobs. Specialization of labor, which greatly reduced the number of people employed in food production, is generally recognized as cultural advancement, but such steps forward are not without psychic and ecological loss. When Carl Jung asserted that "he who is rooted in the soil endures," he was telling us not to forget that we are dependent on nature for our survival, whether we live in a skyscraper or on a farm.

Around the world there are more vegetarians than ever; perhaps as much as half the human population does not eat meat. Antihunting advocates cite this statistic but forget to mention that many of the world's vegetarians come to their diet because there is a shortage of animal protein in their land. Humans choose their diet for many reasons—taste, appearance, weight control, body building, economics, religious laws, personal attitudes, and nutrition. Some people seem to do quite well eating only vegetables, just as some animals are vegetarians. Many other people, however, become ill and some even die following a strict vegetarian diet. In a study of schoolchildren in Guatemala who were poor and had diets lacking animal protein, researchers writing in the professional journal *Pediatrics* reported stunted growth and permanent brain damage due to low protein intake.[4] Other possible diseases of low-protein diets include kwashiorkor and marasmus. In my experience as a counselor and teacher among mainstream Americans, I have seen people who took up vegetarianism for ethical or spiritual reasons develop kidney infections, vaginal yeast infections, dizziness, anxiety, fatigue, and physical weakness, all of which ceased when they began to include animal protein in their diets. According to Dr. Walter C. Willett, departmental chair of nutrition at the Harvard School of Public Health, speaking to those of us who have virtually unlimited choice of foods and dietary vitamin and mineral supplements:

> If one is a strict vegetarian, one has to be somewhat more careful about getting a good balance of different foods to maintain good nutrition. And to some extent, having some chicken and fish makes that a little easier.[5]

Psychologically, you are happiest when you eat the diet that is best for your health. This is based on your unique biochemistry, which may be influenced by your ancestral heritage. When we deny our inner nature, for whatever reason, one result will always be anger and hatred, despite how self-righteous we may be. Recently a good friend of mine was hit by a hit-and-run driver. A mutual friend of ours, the noted Native American medicine man Rolling Thunder, called my bedridden friend and told him that he had to eat lots of red meat and green vegetables because his red blood count was low. The next morning my friend's physician came into his hospital room and told him the same thing. Rolling Thunder believes that the right diet for a person is the one that his or her ancestors ate seven generations ago.

If you are a vegetarian and healthy, more power to you. I am like the cougar; I need to eat meat. I once tried a vegetarian diet for nearly a year; then the East Indian holy man I was studying with at the time told me to stop, because it was running against my nature and making me ill. He was right. I was having trouble sleeping and felt weak. Within a week after returning meat to my diet, my symptoms were nearly gone.

Research shows that many teenagers today are choosing vegetarian diets. Their choice is not based on nutrition but on politics and ethics. In some, this is a sign of assertiveness. In others, it represents a desire to remove oneself from the more basic issues of life, and even a feeling of guilt about being alive. For many anorexics, the overwhelming common psychological issues usually center on guilt, which becomes like a knife in the stomach, poisoning any chance for physical pleasure in life. This in turn leads to the desire to be released from suffering and freed of all earth-bound reality, which then is translated into malnutrition and eventual starvation, often with the claim of seeking spiritual enlightenment. A friend met a spiritual teacher of the East two years ago and shortly thereafter declared he was going to be a vegetarian to demonstrate his sincerity. At dinner recently he ordered steak. I asked him what had made him change his mind. His toenails had fallen off in the last year, he said quietly, and his doctor said it was due to his meatless diet.

The noted Jungian analyst Marie-Louise von Franz has commented on the alienation of modern society, "A world in which nothing on the harsh side is ever allowed is not on the side of life."[6] Whether you eat meat or not, you must have living things killed to feed you. Plants are alive and have feelings, too. And if I do not kill at least some of what I eat, can I ever truly appreciate its value in the great golden web of life? "First God created fools, that was for practice. Then he created vegetarians," Mark Twain once said. Twain loved to bait people, as does rock-and-roll bow hunter Ted Nugent, who some animal rights activists consider Satan himself for his outspoken advocacy of blood sports. With his typical Nugent in-your-face candor, Ted challenges: "If every person went out and killed a chicken each year, the world would be a lot more peaceful place."

Life without hunting is possible. Hunting is a privilege, not a constitutional right. In the near future we will be able to give up the physical act of sex for reproduction and replace it with sperm banks, test tube fertilization, and surrogate motherhood. A world without hunting or sex is possible, and seemingly safer, but is it desirable?

Before you can attempt to answer this question you must first be able to understand the underlying motives for hunting. There are already many hunting books that focus on technique, wildlife biology, and stories of hunting, but none say much about what goes on in the hunter's mind. Reading stories about hunting, one can grasp the excitement of the chase and appreciate the satisfaction of putting food on the table by your own hands. But hunting can be undertaken for a wide variety of motivations. To stereotype anyone's motivations is a cardinal sin in psychology. Drawing on my forty-five years of hunting experiences, aided by college degrees in conservation education, resource planning and conservation, and environmental psychology, ten years of work as a psychotherapist and another ten as a teacher of therapists, and some twenty years of cross-cultural work and research among Native Americans, Eskimo, Polynesians, and Asians, I will describe what I have learned about what goes on the mind of the hunter. Based on these experiences and on a lot of time spent soul-searching, I believe that if you can fully grasp the reasons why people hunt, you will better understand the human species and have clearer ideas about how to create ecological harmony and peace in the world.

In one of the few sociological studies of hunting motivation, Dr. Stephen Kellert of Yale University describes three basic kinds of hunters:[7]

1. Meat hunters—45.5 percent of the hunters say they are primarily out in the woods and marshes to obtain meat for the table. Aside from subsistence hunters, like Eskimo in Alaska, these folks tend to be retired and welcome deer, elk, antelope, or rabbit in their menu. Poachers who hunt for food would have to be included in this category. Growing up in Michigan, I knew a number of people who did woods work like cutting scrub trees for pulp

and firewood. They lived a marginal existence, but they always managed to have food on the table because they always carried a rifle, a fishing rod, and maybe a spear in their pickup truck.

How the necessity of hunting for food can lead to a deeper appreciation of nature and the spirit of the land is beautifully conveyed by noted actor James Earl Jones in his autobiography, *Voices and Silences.* In describing his destitute upbringing on a farm in northern Michigan:

> Because we had to be hunters, Papa introduced us early to firearms. The game helped to feed us. There were always firearms in the house, and we did not need to be Boy Scouts because our daily life trained us in the way of the woods, in survival, in pitting our skill against and with nature. Papa gave me my instruction in the tools needed to hunt and to work the land, and instilled in me a respect for the intricate rhythms of the natural world.[8]

2. Recreational hunters—38.5 percent of the hunters say they hunt as a sport and a hobby, to release tension and have fun. Within this group, motivations range from wanting to spend some time outdoors, to people who use animals to test their marksmanship. Within this group are people for whom hunting is a pastime interchangeable with golf, tennis, barhopping, and gambling in Reno. Within this group lies a small minority of so-called slob hunters, whose behavior gives hunting an undeserved bad reputation and fuels stereotyping by antihunters such as Bill Gilbert who once wrote in *The Saturday Evening Post:*

> The average hunter is a hypocritical nuisance. . . . They have little desire to search for game, but a great desire to kill something that can be tied to a fender or held up in a bar room. . . . Hunters are noisy, belligerent, and the dirtiest of all outdoor-users, littering the landscape with bottles, corn plasters and aspirin tins. They are also dangerous.[9]

3. Nature hunters—17 percent of the hunters Dr. Kellert interviewed said that they enjoy being outside and have a deep "affection, respect, and reverence" for nature. When asked, most will admit that they feel that hunting is much more than a sport. To them, ultimately it is a sacred act with as much as or more meaning than participation in organized religion.

The first two types of hunters are understandable to most people. Their hunting motives are up-front and pragmatic. But nature hunters are a paradoxical mystery to many people. The nature hunter is someone whose deep, reverential feelings for nature are best described, ultimately, as love, and then they turn around and kill what they love. Many people interpret such behavior as sadistic or even psychopathological. Dr. Karl Menninger, the noted humanistic psychiatrist, once declared hunters to be sadists. As a hunter and a psychologist, I think he was dead wrong. In his later years Menninger came to agree with me, as we will learn later. In these times, when nature suffers from our indifference to it, and when countless humans suffer alienation from self and from nature, we need to better understand anyone who claims a deep, abiding love for nature and is willing to commit time, money and energy to conservation. Even if you are not a hunter, if you can read this book with an open mind you will come to know why nature hunters can say that for them, hunting is a spiritual act of love.

In reading this book I hope you will gain a deeper appreciation for nature, as well as understand the motives of at least some hunters. I also want to urge people, both for and against hunting, to have the common sense to go beyond creating quick stereotypes about one another that lead to polarization, violence, even murder. As far as I know, no hunter has yet killed an animal rights activist for harassing him, but if hunters are continually attacked by protesters, especially ones with weapons, war will break out in the woods. Late at night in the deer camps more than one hunter has said that if you want to kill someone, do it in hunting season because you can always say it was an accident. As a researcher of both human and animal behavior, I believe that if people can fully grasp the issues raised by hunting and apply this information to their lives, the world will be a much more sane and peaceful place.

As the hunter's moon rides across the autumn sky, the pulses of nature quicken, and so too do the souls of humans, even among those who do not hunt. What is right for one time, place, and person may not be appropriate for another. The primal energies of the hunt live within us, as well as in the natural world around us, and they can be expressed in many ways. Our personal challenge is to learn to do the right thing for us to be whole, individually and as a society. In learning how to respond to the magic and mystery of hunting in modern times we may find important keys to happiness, health, and peace, as well as ecological balance and proper relationship among species. The hunt can be a great teacher of much more than just the techniques of killing.

When we reach the point in the journey
of human spirit where we want to become
conscious, we must be willing to become real
rather than "good."
SAM KEEN[10]

One

The Sacred Animal

Neither in body nor in mind do we inhabit the
world of those hunting races of the Paleolithic
era, to whose lives and life ways we nevertheless
owe the very forms of our bodies and structures
of our minds. Memories of their animal envoys
still must sleep, somehow, within us; for they
wake a little and stir when we venture into
wilderness.

JOSEPH CAMPBELL[1]

On a normal business day, the license plates on the cars out-side the Vermont office of Dr. Michael Billig are from all around New England, Ontario, and Quebec. Dr. Billig is a chiropractor, and his healing hands and techniques bring him many, many patients. Entering Dr. Billig's office, one is greeted by a gallery of extraordinary art by some of America's foremost wildlife artists tracing back to John James Audubon. The paintings are hung on walls covered by wallpaper featuring images of blue-winged teal ducks. Perched on cabinets and shelves all through the office are lifelike duck sculptures and decoys. If it were not for Dr. Billig's medical certificates, one might wonder if this was a nature center rather than a healer's office.

Billig, a robust man in his early fifties, has a strong handshake and a warm smile. He is a master healer and has a special fascination for waterfowl. He is also a passionate hunter.

Over a spaghetti lunch at his farm, with his Chesapeake Bay retriever sitting at his feet wagging its tail in agreement, we swapped hunting stories, building trust. Today, you cannot always be sure that people will want to talk about hunting the way they once did or that if you say you are a hunter you will be welcomed. For many years, Mike has been on the national board of directors for Ducks Unlimited, an unpaid position. Mike and other DU directors spend considerable amounts of travel time and their own money to help preserve and restore wetlands and the ducks and geese that depend on them. For the passionate hunter who is willing to fall in love with the creatures that are hunted, the desire to give something back to nature bears equal passion to the hunt.

While Mike enjoys hunting in general, he says there is a special place in his soul for hunting eider ducks in Maine: "The last two days of the season, each year, I go over to the Maine coast. It's cold, snowy, icy, but it's the high point of my year." This is not the kind of hunting that would appeal to the neophyte. Temperatures hover around freezing and dip lower as the north wind pushes down the wind-chill factor. If you fell into the Atlantic's icy salt water in December along the Maine coast, hypothermia would set in in seconds. Blustery gale-force winds bring in sheets of snow and hail, but with clocklike regularity, so too come the brightly colored sea ducks, the eider being the most gaudy.

As he enjoys the satisfaction of healing a seriously ill patient, Mike Billig similarly enjoys eider hunting despite the weather, and in recent years despite the attacks of animal rights activists. A year ago his tires were slashed while he was out in the marsh. His love of duck hunting is not a secret, which sometimes brings hate mail from animal rights people. After spending most of an afternoon with Mike, I asked him why he has such a passion for hunting eider ducks. "I don't really know," he said slowly, "at least in terms that I could put into words easily, but you know when it comes down to the bottom line, I would say that those last two days of the season are a religious experience for me."

In support of this, Mike shows me his guns and decoys and then spins yarns about when he has been out there on the rocks with sheets of snow driving across the water, watching for the chunky black-and-white eiders to come to his decoys. Pictures of eiders and eider decoys are everywhere around his home.

I have dropped in without much notice and can't stay long. Mike says that if I'd given him a little warning, he would have cooked an eider for me. "That's a special dish, you understand," he says with a chuckle, but I can tell that he really means it. You can only shoot a few eider in two days. Such hunting is a ritual; it has little connection to hunting for sport or survival. In the Billig home, to eat an eider is to taste the sacrament.

Finally I ask: "Mike, you're a healer, someone who impresses me as having the kind of 'reverence for life' attitude that Albert Schweitzer had. How can you kill animals in good conscience, especially those for which you have so much fondness, like the eiders?"

Mike answers quickly: "I help people get strong enough to kill their diseases. Antibiotics kill germs. A surgeon's knife kills as well as heals. Healers are killers."

This is true. A good healer must have the razor-sharp mind of the killer to destroy pathology whether it is in the mind, the body or the soul. But then Mike is quiet for a minute. He has spotted a woodchuck at the far end of the pasture, and Mike lets the dog out to chase it away so it won't dig holes that his horse could fall into. "We have a bargain, that woodchuck and me," Mike laughs. "He stays out of my pasture, and I leave him alone." Then he sinks back into his chair, pauses for a moment, and says with conviction: "I treat the ducks with the same attitude I'd like to be treated with if I was one of them. I won't shoot unless I know I can kill them. The dog makes sure that wounded birds don't get away. When I kill them, I want it to be as quickly as possible. I don't like suffering."

AMONG THE MANY HUNTERS I've talked with, there is agreement with Mike Billig's feelings about the ultimate spirituality of hunting and the sacredness of animals. These sentiments are common among hunter-gatherer cultures, but are seldom discussed among modern hunters. Animal rights activists view hunting as an act of savage brutality and hunters as psychopathic sadists who torture innocent animals. How a hunter can come to see hunting as a spiritual practice and the animals hunted as sacred is a psychology that may seem fraught with contradictions, but with an open mind and patience, the validity of these views can be understood and the truth becomes as clear as the sky on a crisp October day.

The Sacred Animals of the Hunt

Growing up in the marshes of Lake Erie, with ducks and fish and the wind as my brothers and sisters, I was witness to millions of migrating waterfowl every year. If you looked at my notes from my tenth grade world-history class, which had a window looking out over the Detroit River and the marshes on the Canadian side, you would have thought the class was one in wildlife art. To this day, whenever I see a flock of ducks or geese take to the air, I feel compelled to stop whatever I'm doing and watch. The larger the flock, the greater the reverence I feel.

Providing that a person is not afraid of nature and does not feel threatened by untamed wilderness, there is a natural attraction between the mind and nature. The symbols of our unconscious, which we meet in our dreams, are the same as the things we see consciously in the natural world. When two similar symbols come into harmony, energy is exchanged and both are enhanced by the harmonic bond. This is why a beautiful scene like a sunrise, a waterfall, a snow-capped mountain, a towering cedar tree, or a flock of migrating geese can enchant us so easily. Studies of elderly people around the world who take a zest for life into their golden years consistently show that the ability to appreciate beauty in nature is one of the wellsprings of their longevity.

The element of serendipity also adds to the expectant excitement of watching wild animals. You cannot predict what you will see, hear or encounter when you depart from the buildings that shield us from the natural world. While the general forms and features of nature are by themselves comforting and inspiring, it is the possibility of something extraordinary happening that adds excitement to any venture into the natural world.

One of the skills a hunter must learn is to call animals, which for me meant that my first "musical instrument" to master was a duck call. When I was a teenager, I would study duck language by sneaking up on feeding ducks at night, when they could not see me, and then attempt to engage them in conversation with a call. When I was about sixteen, one cold November night, three friends and I went to practice calling ducks. Approaching the river, we crawled through a hundred yards of willow thickets before slipping out onto a sandy beach on the western shore of the island, where thousands of ducks could be heard quacking and splashing as they fed in the safety of darkness. Lights from the village of Gibraltar, a mile across the river to the west, sent reflecting beams across the water, and we could see black shapes swimming close to the shore.

We crawled into the roots of a fallen willow tree and began to converse with the ducks. We called and the ducks responded. Soon several dozen were within fifty feet of us. But then we heard the excited barking of my friend Roger's pet beagle. "The dog got out," he moaned. We hoped he wouldn't follow our trail and blow our cover.

The dog was a quarter of a mile away when suddenly his bark changed to an excited howl. He was after a rabbit. For the next five minutes we had two dramas going on: the ducks in the river at our feet, and the dog chasing a rabbit in the marsh behind us. As the dog's barking got closer and closer, the ducks that had come so close to us began to mill around nervously. Now the dog was less than fifty yards away. Suddenly Roger fell over backward, holding his stomach, screaming, "I got him! I got him!" Ducks seemed to explode everywhere, sending a shower of water on us, and the dog came flying into our midst, barking like an animal possessed.

What Roger had caught, we soon found, was the rabbit! He had borrowed his father's warm hunting jacket, which had a button missing on the front. Hunching down among the willow roots, the front of the jacket had popped open, creating what must have looked like an ideal hole for the rabbit to escape into. In the heat of the chase, the rabbit had simply jumped into his jacket.

Roger just left the rabbit in his coat and we walked home. We knew that no one would believe what had happened, so we showed the rabbit to my parents. Then we got in the car, put the rabbit into the glove compartment, and made the rounds to show everyone else. We could not allow our story to be dismissed as fantasy. Finally, we let the rabbit go.

Experiences like this show why watching wildlife is rapidly becoming one of the nation's most popular pastimes. In 1991, while there were 14.1 million hunters, 76.1 million others said they participated in recreational activities involving wildlife watching. For many people, binoculars, sketch pad, and camera have replaced guns as the tools taken into the field. In an age when modern living pulls us away from nature, it is critical that as many people as possible surrender to the desire to watch wildlife, for the rewards cannot be equaled by watching television or reading a magazine. For those who choose to hunt, however, the experience of nature kinship offers possibilities for deeper meaning.

While fishing and hunting share the common quest for capturing a wild creature, hunting for me has always had a more seductive call. A fisherman waits, seldom seeing what he hopes to catch, and cultivates patience and perhaps skill in casting. Once a fish is hooked, excitement rises to be sure, but once the fish is landed it can be returned to the water to live on. Also, relatively few fish that get off the line before being landed are harmed or killed by being hooked. There is more leniency in fishing. A hunter holds life and death in his hands, with creatures for which we have a closer kinship.

Once a hunter decides to shoot a gun or release an arrow, there is no question of catch and release. If successful, the result of the hunt is final for the hunted. The responsibility for taking another life adds a dimension to hunting that does not exist for the nonconsumptive recreationist. As Dudley Young observes in *The Origins of the Sacred*:

What is religious about hunting is that it leads us to remember and accept the violent nature of our condition,
that every animal that eats will in turn one day be eaten.
The hunt keeps us honest.[2]

Sitting in a duck blind on an icy November morning, watching
the sunrise as strings of ducks and geese rise up, walking across a
golden field of ripe grain in search of a pheasant to flush, or stalking deer with a bow and arrow, the hunter feels an incredible
sense of excitement. Then comes the flush of the pheasant or the
explosion of the rabbit from the brushpile, and the hunter's reaction: Surging adrenaline. Chance. Skill to be expressed. And some
tasty meat to take home. Ultimately, satisfaction. Hunting can be
a diversion or an escape, but if it could be simply substituted by a
nonconsumptive use, such as bird watching or nature photogra-
phy, very few would continue to hunt, for it is quite an expensive
activity. For a weekend of hunting ducks and geese in California, a
hunter must spend more than fifty dollars for a resident license
(nonresident licenses are considerably higher) and the necessary
state and federal stamps. Steel shot, now required for waterfowl to
prevent birds from getting lead poisoning (caused when they ingest spent lead pellets they find at the bottom of the marsh) costs
at least ten dollars a box. Reliable shotguns start at about two
hundred dollars. Decoys, waders, boat, motor, clothing, gloves,
plus gas, food, lodging, and perhaps a guide, quickly make hunting
a very costly venture. There must be more to hunting than just
watching nature, otherwise hunters would not spend the money
they do. And if gambling is the real motive driving hunters, they
would probably get greater returns from the casinos at Reno or
Las Vegas.

The chase is the primary focus in some types of hunting. Pursuing rabbits, deer, bears, or raccoons with hunting dogs is like a
foot race with a life-or-death reward for the winner. Hunting fox
with hounds involves the additional challenge of horsemanship,
which adds more of an element of risk. Research on stimulus-
seeking athletes such as sky divers, white-water kayakers, Bungee

jumpers, and mountain climbers has shown that they are not necessarily wackos; rather, most are well-balanced individuals who have a special craving for the adrenaline boost that results from taking risks. The risks they take serve as a way to integrate consciousness, and the adrenaline rush from risk-taking and mastering challenges gives them high self-esteem. Generally, they do not seek drugs for highs. As I discussed in my book *Nature as Teacher and Healer*, when drug addicts are led into high-risk outdoor sports they become cured of their addiction at a rate higher than nearly all other treatment modalities. Risk-taking builds self-esteem, which seems the best medicine for addicts.

In some cases, the act of killing can take on certain aspects of combat; the more handicapped the hunter, the more excitement possible. There is more of an energy rush when you tackle a lion with a spear than when you drop one at four hundred yards with a 30-06. If you savor opportunities to challenge death, consider the emotions experienced in the traditional Russian style of hunting wolves. As described in George Bird Grinnell and Theodore Roosevelt's classic *Hunting In Many Lands*, published in 1895, wolf-coursing recruited entire village populations as beaters to flush out wolves that preyed on local livestock. When a wolf was sighted, the hunters would release massive Russian wolfhounds, and the chase would begin. Upon cornering a wolf, the hunter would wade into the frenzied pack of giant dogs surrounding the wolf and, with a special hand knife, kill the wolf by slitting its throat or stabbing it in the heart. Participating in such a sport could lead some people into ecstatic states of mind not unlike those attained during religious practices such as dervish dancing or walking barefoot across burning coals, or those moments of spiritual communion that some soldiers experience in the heat of battle. It is little wonder, then, that in reporting on his experiences of wolf-coursing in the 1800s Roger Williams wrote: "The enjoyment of [the] sport increases in proportion to the amount of danger to man and beast engaged in it."[3]

While there are some similarities between hunting and warfare, there is a significant difference between the psychology of the warrior and that of the hunter. The warrior engages in battle against something to protect others, often risking his own life on behalf of others'. The warrior must conquer fear of personal injury and death, as well as objection to killing another human. Military leaders use various motivational psychologies to help the soldier kill, the most common of which is to portray the enemy as evil and to make the act of killing mechanical and thus without emotion. The political and military rewards of winning are safety and security. The personal rewards for the warriors come from serving a cause and demonstrating courage. The sentiments of having been a warrior are frequently preserved by displaying souvenirs, medals, and other keepsakes, and sometimes trophies of battle, such as enemy flags and weapons.

The modern hunter, on the other hand, is challenged not so much by fear as by overcoming guilt. Most animals killed by hunters are not taken for protection or self-defense, but for food, and perhaps for a trophy. There is a special fondness in our hearts for wild things, and a hunter must work through guilt feelings to be successful. The more one learns about wild animals, the more one develops a fondness for them. Hunters also enjoy the overall act of hunting; being out in nature itself is a pleasure. When they are successful, it is customary for hunters to share their kill with others, whether it is in a tribal society or a modern big-game dinner. And when the season is over, hunters often spend many hours helping to preserve habitats and restore species. The emotions that good hunters need to cultivate are love and service more than courage. The sentiments of the hunt then become translated into art, as much as or more than trophies. Even the trophies of the hunter are art, for good taxidermy requires extraordinary artistic skill.

I understand the excitement of the chase. When I was younger I owned a black-and-tan coonhound named Zip that led us for many extraordinary chases through the swamp forests along the Lake Erie shore. But with time, I have come to prefer a more passive style of hunting. My choice to sit or stalk rather than race after animals is not due to the physical demands but rather to my preference of the mental states that come from waiting and blending into nature to become a witness. Sitting in a duck blind with a howling north wind at your back, or perched on a stump in hopes of the appearance of a deer on a snowy December day, is a cleansing meditation of the highest order. After months, perhaps years, of preparation, the rest of your life seems to slip away, and the gestalt gently comes together as the decoys are set and you retreat to the blind as the sun rises. Unfinished business of the moment washes away with the wind and waves, and each sound, odor, and emerging form in the sky or among the trees becomes a potential trigger for excitement. Senses we do not normally use are brought into action. We rediscover what hope and faith feel like while waiting for game to appear, and as a result we remember how good it can feel to be alive.

One of the first things a successful hunter must learn is to see. Because we have become so dependent on the printed word and scanning television sets and magazines for orienting to the world, many people today have never learned to develop perceptual skills such as peripheral vision, acuity to sounds, and seeing patterns. Spotting an animal before it sees you is one of the most crucial skills a hunter must develop. With time, you learn to let go of the need to see detail and let your overall field of vision become a soft, blurred tapestry of forms and shapes as you scan landscapes. What you learn to look and listen for are the anomalies: the branch that moves and becomes a leg or antlers, the acorn that drops on dry leaves from the paws of a feeding squirrel, the black speck on the horizon whose regular rapid wing beats and streamlined body shape tell you it is a duck or goose and not a soaring vulture on still wings, a sea gull with a slow flap and glide, or the irregular wing beats of a crow. See the whole, look for the variations in patterns, and then focus in for detail; these are the essential perceptual skills for a successful hunter. When you suddenly switch from seeing patterns to looking at details, seeking what Zen meditators call "one-pointedness of mind," you have the greatest chance for accuracy.

In hunting, timing is crucial. It is little wonder that hunters eventually become expert in the perception of time. The deeper meanings of the linear functions of time are realized by the hunter in the experience of taking the life of another living thing. On some level, when you stand there with a cock ring-necked pheasant or a bull elk that you've just shot, you cannot help but acknowledge that like the animal whose life has just been taken, you too will die someday.

The annual procession of seasons, and the ways in which animals must respond to them to survive—migrations, hibernations, mating frenzies, or nesting swarms—remind us of the cyclical function of time, which balances the linear progression of hours, minutes, days, and years that govern our schedules. Each year a procession of species of ducks pass through the northern Lake Erie marshes—first the tiny teal, then mallards and early black ducks, then gadwall and widgeon, then diving ducks, first bluebill and redhead, then canvasback, and finally goldeneye and Canadian red-legged black ducks come with the snow storms of December. Old hunters at Pointe Mouillee marsh along the shores of Lake Erie say they can set their watches by the time the scaup or bluebill ducks arrive each fall. With clockwork regularity, on October 25, the first flocks of these black-and-white diving ducks with bright blue bills come floating down out of the northern skies, forerunners of the immense rafts of hundreds of thousands of bluebills that will pass through in the next few weeks. Coupling these two forces of time together, the linear and the circular, you come to recognize that time is like a child's slinky toy.

On top of the slinky form of time comes human consciousness, which adds the dimension of ever-growing learning. Year after year, returning to the same blind or stand with a little more knowledge than the year before, the hunter's time becomes a spiral as the experiences of the past become integrated into an enriched sense of nature in the present. And as wisdom grows, one learns that in nature, occasionally, windows open and we can glimpse patterns that transcend temporal time; moments of startling awareness that occur when past, present, and future exist at the same time. Among the Inuit or Eskimo of the Arctic, who still live by subsistence hunting and fishing, life is normally structured in hopes of catching glimpses of the future more frequently—it is called "seeing." In a land where harsh weather can sweep away life in a second, and loss of touch with the animals may mean famine, one must try to cultivate the ability to anticipate the future as a survival skill. A seasoned hunter also knows, as his ancestors knew, that occasionally animals can be messengers.

The Indian mounds at Azatlan in Wisconsin draw thousands of tourists every year to contemplate how life must have been during the times of the Mississippian mound builders one thousand years or more ago. A ranger at Azatlan recently related a tale that raises the question of the extent of the wisdom of wild creatures. There were plans, it seems, to build a log cabin next to a burial mound known as the Princess Mound. As the workmen laid out the foundation for the cabin, a snow-white albino robin suddenly appeared. The laborers, all Native Americans, refused to go ahead with the construction as they felt that the robin was an omen, perhaps even the spirit of the people buried in the mound expressing their protest about building so close to a sacred burial site. The robin stayed around the mound all summer long. By fall, it was decided to move the cabin elsewhere. As soon as the decision was made, the robin left and was never seen again.

There is a very definite tao of hunting. The skills of the adept hunter are very similar to those of the martial arts masters.

One of the most important lessons in mastering a martial art is that of feeling a sense of oneness not only with yourself but with your opponent. In hunting, the same is true. A good hunter must learn to link personal thoughts, actions, and moods with the larger forces of nature to increase his chances for success. Being at the right place at the right time is at the core of hunting success. A hunter seeks to blend with the overall field of mind within which we live, putting aside personal ego to increase the intuitive sensing needed to pull out harmonies from the collective consciousness.

The nature hunter, who takes the time to learn to work with nature's way, must develop an acute sympathy with the animals he hunts. He must not only have a good deal of knowledge about them, he must have a feeling for them, which is a reflection of how he views himself. The path of the nature hunter then leads him not toward violence and mayhem but toward respect, awe, humility, and even love for the animals hunted.

There are many conditions in nature that collectively lead the hunter's mind into a state of awe and reverence, which is the emotional foundation for transcendence. The buck appears out of the swamp, or the flock of honking Canada geese passing overhead veer and set their wings above the decoys thirty-five yards away. Heart pounding wildly, eyes wide, pupils constricted, mouth dry— all your senses are peaked. Your hands shake as you slip off the safety, adrenaline is surging through your body, and then in a split second you have shot and, you hope, cleanly killed.

In hunting, the kill is a potential peak experience. Compare the following two statements, the first by William James speaking of the religious experience, and the second by modern hunter Ted Nugent describing his feelings at the moment of killing a nine-point Texas white-tailed buck:

> For the moment nothing but an ineffable joy and exaltation remained. It is impossible fully to describe the experience. It was like the effect of some great orchestra, when all the separate notes have melted into one swelling harmony, that leaves the listener conscious of nothing save that his soul is wafted upwards and almost bursting with emotion.[4]

> This is my magic time. I had not taken this majestic beast. He was given to me. I am merely a piece of the puzzle. There was no negative. No wars. No hate. No taxes. No

pain. I am truly moved at these moments. I really don't understand them completely. I know I am participating in life. It's a gift. This is how we are made. This is truth. Sustenance. I am at once exhilarated and at peace. Movements seem slow motion.[5]

As a psychologist, for more than two decades I have studied the experiences of spirituality. Manifestations of the sacred are not always angelic. They can be precipitated by many things including accidents, near-death experiences, meditation, yoga, martial arts, sex, childbirth, drugs, fasting, prayer, physical traumas, religious rituals, and psychotic episodes. Sports also can bring about spiritual states of mind, and while I have seen many people in chaos upon entering an unexpected spiritual state, I have never seen an athlete have a bad trip, like that caused by a drug overdose. According to Michael Murphy, perhaps the foremost researcher of consciousness and physical activity of our times, common qualities of the peak experience in sports are:

1. extraordinary perceptual clarity
2. extreme focused concentration—one-pointedness of mind
3. a sense of emptiness and a feeling of unity that seem to envelop the field of play, resulting in a feeling that time itself has slowed down and one's actions have become an expression of merger with a higher force or purpose
4. access to new energies which fill one's mind and body
5. access to experiences which may fall within the paranormal[6]

Many hunters will recognize these qualities in the experience of hunting. Hunting can be a path to spirit if you can surrender to its power and become proficient in the act of executing the kill. Some people may charge that killing animals to get "high" is egotism at its worst, self-indulgence, and a power trip without compassion. These charges can only be leveled by someone who has never honestly hunted. Often, people who make such statements present themselves as being "spiritual." Frankly, I think they would be more accurate to say they are self-righteous. A truly spiritual person does not judge others if they are following an honest path of the heart, and among the paths to spirit, there is that of the hunter. Gandhi said: "Everyone should follow his own inner voice."

Hunting may lead people to have peak experiences. All the positive elements are there, from spectacular environmental settings to intense emotional excitement, to encounters with the deepest issues of life and death. Many hunters I know feel that ultimately hunting is their religion, but often they do not admit this because of criticism from those who do not understand the hunter's soul.

And what about the poor animals? the critics of hunting scream. Anyone can declare an animal to be special, even sacred. But a thing can become truly sacred only if a person knows in his or her heart that the object or creature can somehow serve as a conduit to a realm of existence that transcends the temporal. If hunting can be a path to spirit, unhindered by guilt, then nature has a way of making sure that hunters feel compassion. "For those who have a religious experience all nature is capable of revealing itself as cosmic sacrality," concludes religious historian Mircea Eliade in his monumental review of spirituality, *The Sacred and the Profane*.[7] It is precisely for this reason that Theodore Roosevelt, one of the greatest hunters and friends of wildlife in the twentieth century, often proclaimed that "all hunters should be nature lovers."

What we can say from the decades of in-depth psychology research is that human nature is neither evil nor hateful in its original state. These conditions are the results of fear embedded into the self, which is the cause of all emotional imbalance. In his studies of extraordinarily healthy people whom he called "self-actualizers," psychologist Abraham Maslow found that a common quality of self-actualized people was a love for nature. And so with this in mind, let us turn to the animals that are the other half of the drama of the hunt. How can a hunter come to feel that an animal is sacred?

If you do not try to outrun your quarry, success in hunting is a matter of synchronicity as much as skill. People talk about luck. I believe people make their luck happen out of clear intention and self-knowledge. A hunter may know the habits of the animals, and he may erect concealment blinds, use calls, scents, and decoys in hopes of attracting the quarry, and suit up in state-of-the-art camouflage clothing, but he cannot predict how wild animals will behave. More than one hunter has been intently looking for the deer to come down a path, only to be scared out of their wits by a snort from just a few feet behind from the buck who noiselessly slipped up to investigate the intruder. The magic of nature involves the sympathetic attraction of two or more elements. You cannot control the animals you hunt, but you can interact with them on many levels. Dimensions beyond our control help bring together hunter and hunted, I am convinced.

"Civilized" people often scoff at the idea shared by native peoples around the world that animals may serve as messengers and even guides to the future. Many a woodsman will predict the severity of the coming winter based on the width of the brown midsection of a woolly bear caterpillar or the heaviness of a fox squirrel's fall coat. Close observation of animal behavior suggests that animals may be much more in touch with the web of life than we think—they do more than just fill their niche in the food chain. Watching swarms of blackbirds working through a marsh, clouds of starlings coming to roost at nightfall, flocks of geese automatically changing formations, with almost mathematical precision, and schools of fish moving as one organism, one cannot help but wonder how they orient so perfectly. According to Japanese scholar Kazuko Tsurumi of Sophia University, such behavior is "collective intuition" at work. In the paintings and carvings of Pacific Northwest Coast Indians, which frequently depict animals with eyes all over their bodies, the eyes tell us of how animals know how to live in harmony with the larger dimensions of time and let the forces of nature guide them with wisdom that goes beyond the five senses. They learn to perceive with their entire bodies as a sensory organ. For this reason, native peoples speak of the "wise animals" and pay special attention to unusual animal behavior or appearances that may indicate future events or impart messages from the spirit world.

The wisdom of native peoples, which we will explore in more detail in chapter 2, also asserts that under the right conditions, the success of the hunter is not just a reflection of skill but the choice of the animal. All around the world, there is a commonly shared belief that within each species is an ongoing sympathetic web of communication. It is not a loose web, either, but one that is presided over by wise elders. I have seen evidence of this especially in flocks of geese and sandhill cranes. Several hundred birds may be milling in a field, but if you look carefully, off to the side, you will see a small group, perhaps five to ten, slightly larger, and even with an ever-so-slightly different coloration. Some people refer to them as "sentries."

Many people who have closely watched wildlife have seen the guardians or leaders of a flock. According to shamanic psychology, these creatures are more closely linked with the spiritual dimensions of life, just as human priests and shamans serve as a council of elders for our species. Shamans, who were the first game biologists, enter trance states to mentally visit such councils of elders to establish guidelines for the hunt that ensures the survival of both species. If people take only what they need and the species is not endangered, then the spirit keepers are happy and blessings will come to those who follow the shamans' advice. If people do not honor the animals they kill and do not limit their catch according to what the spirit keepers advise, then accidents and ill health may befall the hunters.

Anthropologist Bruce Finson, who has studied extensively with the Huichol Indians of Mexico, reports that on several occasions he has seen deer appear to sacrifice themselves to Huichol hunters after a sacred deer ceremony has been held. In the Huichol culture, a major religious event each year is the pilgrimage to the sacred high desert called *Wirikuta*. The guardian spirit for this trek is *kayumari*, the deer. In preparation for the hunt, the Huichol hunter engages in a period of purification that includes abstaining from certain foods (especially salt), from sex, and from drinking waters from certain sacred springs. During this time they also pray and make offerings to the deer, showing their humility and respect for the animals they ask to kill.

When a Huichol hunter hunts a deer, he does not try to chase it, Finson reports. Instead he looks for a deer that will stand, facing him, and not run away when it sees he is a hunter. The deer "talks" to the Huichol hunter with its eyes, they say. For the hunter to find the willing victim, he must be in the proper state of mind.

When the Huichol hunter kills a deer in a sacred way, he offers prayers to *Tatewari*, or God, as well as *Maxa Kwaxi*, the Elder Brother Deer. After the flesh, hide, and other valuable parts have been taken, the bones are ceremonially buried in a forest so that the deer may regrow from his bones.

The modern hunter does not have the cultural system of the Huichol, but every hunter knows those certain people who seem to have a special calling for a particular species of bird or mammal. Mike Billig has his eider ducks. Ted Nugent is a deer man. I know people who have a special knack for hunting bear, caribou, raccoon, canvasback ducks, pintail, elk, quail, and wild pigs. As we will see in chapter 2, these kinds of interspecies ties moved native peoples to develop elaborate cultural systems of stories, songs, rituals, and dances to honor the animals of the hunt, often using them to preserve cultural mores and ethics.

In modern times, invariably those fortunate hunters who come to know their animal kin are the ones who go on to found organizations to support the preservation of the species they hunt. In the final analysis, the emotion that drives these people is not guilt, as some might like you to believe, but love. At the root of any true feelings of sacredness there must be love, for it is the core emotion of self-value, which is the ultimate survival motivation of our species.

The Animal as Teacher

I have not been able to take up Mike Billig on his offer to hunt eiders with him, but I have come to understand how a person can see a single species in a special way through a relationship that I have with snow geese. While the Great Lakes attract millions of ducks and geese every year, in some twenty years of living and hunting in Michigan, only once did I see snow geese close up, and when I did, the sight left a permanent impression in my mind. It was an early November morning and I and some friends were sitting in a duck blind of cattails and willows, watching the sun rise over Canada several miles away across the cold, gray water. A low fog hovered about fifty feet above the water. Out in the center of the bay, we could hear the cries of a large raft of feeding waterfowl, but we could not see them. Slowly the sun burned through the fog, reducing the birds' protective cover. As the flock became visible, birds began to take off and head for the sanctuary of the open waters of Lake Erie. On mornings like this the excitement of the hunt blends into the chill of the morning, and the chills that run down your spine are pronounced. On this morning the emotion of the moment took on a peculiar, excited softness when suddenly from the fog came some strange cries, a new call, higher pitched than the honking Canada geese, a cross between a musical note and an almost human voice. The sounds seemed to cast a spell, announcing that something out of the ordinary was about to happen. Then, out of the fog, some fifty feet above the water and less than a hundred yards away, came a line of eight snow geese, pure white, with black wing tips and pink bills and feet. They appeared just as the golden globe of the sun broke above the horizon. None of us could raise a gun to shoot. Like a fresh, clean wind, they passed over us as we sat mesmerized by their presence. I sensed that there was something very special about the moment, but I had no idea what it could be.

I did not even begin to comprehend the meaning of that moment until nearly fifteen years later. It was August 1973, and I was in Alaska. I had been asked by a major environmental organization to deliver a paper on the nature of public criticism of environmental groups at their fall convention. I had wanted to do some original research for this presentation, looking at a case study of a hot issue where there were some strong antienvironmentalist sentiments, and with some luck I was given a research grant to study attitudes of the Alaskans' about the soon-to-be-built trans-Alaskan pipeline. That summer I walked, rode, and flew over the entire length of the proposed pipeline and was finally given a chance to go to the oil field at Prudhoe Bay on the shore of the Arctic Ocean.

Alaska is a land that teaches humility by the vastness of its wilderness. Passing over the majestic snow-capped Brooks Range, as the Lear jet began to descend, the immense carpet of gray and red tundra with numerous pockets of open water and countless serpentine glacier-fed streams loomed up to meet us. On the ground, the first thing that struck me was the flatness of the land. In all directions but south, there was virtually nothing to block the horizon.

In shirtsleeve weather, I walked and drove along gravel roads, inspecting stacks of pipe waiting for the go-ahead for a conduit of black gold stretching south to Valdez, with its year-round ice-free port. Caribou, plentiful like cattle, were grazing everywhere. On several occasions we had to wait for them to move out of the way as we approached. The boggy tundra is not nearly as easy to walk on as the hard-surfaced roads, the caribou had decided, and so, like the cars and trucks, they used the new gravel roads with relish.

As my visit to Prudhoe Bay was about to come to an end, I went off for a walk, to be alone and catch a few last precious minutes in this primal land where so few human feet have touched the ground. The silence, I found, was almost deafening. It made my ears ring as they searched for the familiar background noises of civilization. Then, from out of the great void, like tinkling bells, their calls came. I scanned the sky. Off to the south, I finally saw a thin line of white that was moving ever so slightly. Like phantoms they came, snow geese with wings set, gently gliding down. Then, on cue, they suddenly swooped down and settled on a small island in the middle of the nearby Sagavanirktok River.

I could hear their conversation as they began feeding half a mile away across the flat tundra. There are two species of snow geese in North America. The greater snow goose, *Chen atlantica*, which nests to the east, on Ellesmere Island at the northern limits of the Beaufort Sea, is most likely the species I first saw on Lake Erie. These birds were their cousins, *Chen caerulescens*, the lesser snow goose, which nests farther west on the tundra, extending to Alaska and Siberia. There are at least 3 million snow geese in North America, the largest breeding colonies being along Hudson Bay. The flock before me was part of a Western Flyway annual flight pattern. They would winter in Southern California down to Mexico, then return to Alaska for nesting. One could say there was nothing unusual about such a sight, yet I felt something very special about the presence of these white geese. Living creatures, whether they are geese or humans, embody the spirit of the land where they are born. The environmental fields of that location imprint in their bones as they are waiting to be born, forever giving them a magnetic resonance with their place of birth. They become a spirit of that place.

As had been the case when the snow geese appeared on Lake Erie, my life was in a period of significant transition. In August 1973 I was moving from one university to another, and changing jobs, from being a professor of environmental studies to being a teacher of human services and psychology. Learn to watch for the unusual events in nature and you will find that indeed there are animal guides for each of us, and each has a special kind of wisdom to share.

Unlike some waterfowl, snow geese are doing very well and their survival strategy suggests lessons for life from which we can all benefit. No one can farm the arctic tundra where they breed, and hunting limits and seasons are carefully adjusted every year according to breeding success. Ironically, the reduction in duck populations caused by the loss of wetlands to farms has provided more grain for geese, so goose populations today are rising. The snow geese are survivors, wisely making the best use of their resources without any complaint. They embody wisdom and seem very low on vanity.

In talking with biologists who study the Arctic, I have learned that a nesting colony of snow geese was established on Howe Island in the Sagavanirktok River in 1980. It began with twenty-four birds. Thirteen years later it has grown to over two thousand, and shows signs of increasing, although the arctic foxes and grizzly bears may have a say in that. In mid-May the snow geese settle in to nest. They all nest at the same time, "swarming," as biologists call it, as a protective mechanism. During the time when the females are on the nest, the eggs and young birds are extremely vulnerable, and predators abound. Because there are so many young at the same time it seems that the birds hope that the odds for survival are in their favor. But it doesn't always work. From 1990 through 1992, foxes got onto Howe Island and went on a killing spree, catching all the young nestlings and eating or carrying off and burying all the eggs. Virtually no young snow geese were produced in those years in the Howe Island colony. In 1993, the snow geese escaped the foxes, and so about a thousand young goslings swam off with their parents to the nearby marshes. Brant, a smaller species of Canada goose, also nest on Howe Island, their nesting time immediately following that of the snow geese. In 1991, a mother grizzly and three cubs swam out to Howe Island just as the snow geese were leaving. In short order the four bears decimated the nesting brant, feasting on eggs until none were left. In 1993, foxes got all the brant eggs. When a species with high food value grows, it attracts predators. The fox population in Prudhoe

Bay has jumped dramatically since 1980, aided by the multiplying snow geese in the colony on Howe Island. As the numbers of foxes have risen, so has the incidence of rabies. Nature is self-regulating. As a population reaches the limits of its ecological capacity, if one balancing force doesn't reduce the numbers, another will.

The snow goose nests the farthest north on the continent, keeping its numbers up by massing in large colonies all at one time—letting the odds help at least some get through this time of helplessness, because life for a snow goose is a continual series of confrontations with hunters. Foxes, bears, ravens, jaegers, and glaucous gulls relish snow goose eggs and nestlings. Just after hatching their eggs, the adult birds molt and are flightless for nearly a month as both young and adult birds grow new feathers for the long migration ahead. In the flightless time golden eagles, goshawks, and wolverines may join the other predators hoping for a snow goose dinner. Inuit do not harvest eggs or adult snow geese at Howe Island, but they do elsewhere, and this is a matter of concern to people in other parts of the snow goose's annual migration range.

Snow geese fly high and fast, often taking advantage of high-altitude winds to ease long distance travel. Barring the dangers of eating grain poisoned by pesticides, devastation by predators or by hurricanes, or colliding with electric power lines in fog—and as long as the arctic tundra is preserved and they can find sanctuary in wintering there—snow geese will be with us for a long, long time. Hunters take some, normally from 6 to 8 percent of the flock, which is not enough to pose any threat to the species. Snow geese ride on the north wind, the breath of the Great Bear, whose mythic qualities in native psychology are strength and purity. They also flock in large numbers, exhibiting an unusual ability to organize their species and cooperate with one another, which can be vital to their survival.

You can learn a great deal about nature from studying just one species. Millions of nonhunting wildlife watchers share the chance to learn from animals if their behavior is watched with insight as well as appreciation of wonder and beauty. Yet, watching alone does not fully explain the attraction we feel for wild creatures. If one grasps the principle of collective intuition, one must ultimately ask just how far does the mind field go to which snow geese are tuned in? Any hunter who has spent much time in nature knows that there is some kind of psychic interplay between animals and hunters. Let us suppose for a minute that if in the collective mind it is possible to sense and feel the thoughts and emotions of another living creature, must there not be some advance inkling of the intent to kill in the mind of the hunter, whether man or beast?

The Inuit, who, like the snow geese, have learned to be survivors in a land of harsh climate, say that all species of animals are linked together in an invisible web of consciousness. In the overall scheme of things, the Inuit insist, this consciousness extends to other species, including man. Anyone who has ever felt the mood of a crowd change knows what "collective contagion of thought" means. And when such contagion occurs, unusual things may happen. Dr. Albert Schweitzer is forever remembered for his eloquent writing about the ethics of life and death, especially his concept of "reverence for life." Schweitzer came to this concept with the aid of the animals one day when he found himself in a potentially dangerous situation in the middle of a herd of hippos. We tend to think of the hippo as a gentle, fun-loving blob, but whether in water or at night on land, the hippo can have a nasty disposition and in fact is one of the most unpredictable African animals. Schweitzer considered the hippos among the most dangerous of all African wildlife. Thus, as his boat drifted into the middle of a number of submerged hippos, everyone was on edge while the hippos began to snort menacingly. In the midst of this dangerous situation, Schweitzer had what might be called a "peak experience," where one may grasp a kernel of pure truth. He wrote:

> On the afternoon of the third day, as we were passing through a herd of hippopotamuses, there came to me suddenly unpresaged and unsought, the words "Reverence for Life."[8]

I was first trained as a wildlife biologist, learning to observe and record animal behavior. Later, as I came to realize that all environmental problems begin in the human mind, I became a behavioral scientist and learned to watch people. Both kinds of scientist in me, the biologist and the psychologist, were trained to be very skeptical of anything that does not fit into a pragmatic paradigm. But I also wear other hats, as an entertainer and a producer of concerts and conferences. There I work with the emotions, feelings, and intuitions of "collective contagion" as a participant, not an observer. In 1978, I was asked by some Indian spiritual leaders to produce a modern powwow for Indians and nonIndians. As a result of this work, I came to understand a deeper relationship between people in animals.

The Indians told me that the modern word *powwow* comes from the Algonquin root word *pau-waw*, which means "he who dreams." They said that the best events are ones guided by dreams and they wanted to see if my dreams would give me guidance in producing their event.

They held a dream ceremony, and, to my amazement, I began to have extraordinarily vivid dreams. Some were very mythic and symbolic and others extremely concrete, including specific information on where to hold the event, whom to invite, and even how much to charge. These dreams lasted for nearly a year, the entire time to plan and conduct the gathering, which was very successful. At the end of this time, seemingly as a thank-you for my help, I had an extremely unusual dream. While I was asleep, in bed in Seattle, Washington, I dreamed that I was back on Grosse Ile, standing on the southernmost tip of the island where I had duck hunted during my youth. All my senses were as alive as in full waking consciousness, and time and space seemed condensed into a vital present. It was afternoon of a cold autumn day and the sun was setting in the red western sky. To the east, black snow clouds loomed over the gray water of the lake, where whitecaps frothed with gusts of the west wind. Then out of the dark cloud to the east came some white specks. They looked like snowflakes at first, but as they approached, I saw that they were snow geese. They came closer and closer and then set their wings to land in the calmer waters at my feet. Just as they were about to land, an extraordinary thing happened. They transformed into short, dark-skinned, bare-chested men wearing leather leggings. Their leader approached me. His expression was warm but very serious. In his hand was a white candle. Our eyes met and then he shoved the white candle into my chest. I was engulfed in warmth and light and then I woke up crying. Another turn on the spiral of time had come, heralded by the snow geese dream. What deeper level of meaning did snow geese have for me? I wondered.

Upon sharing this dream with the Indian medicine man Rolling Thunder, he told me I should study the culture of the Lapps or Saami of northern Scandinavia, as there was something very important there. After some research I found that my family name, Swan, was not of Scottish origin as I had always thought, but that the Swans of Scotland were descended from the Sweyns, who came to Scotland from Norway, led by a short, dark-haired chief named Olaf. The snow geese had guided me back to my correct ancestral lineage, a priceless gift in self-awareness.

The word *totem* comes from an Ojibway word that means "my fellow clansman," or "he is my relative." One might call the snow goose my "totem animal." These "dreams of the spirit," as Rolling Thunder called them, were more vivid and dramatic than anything else I had ever experienced.

According to Jungian analyst Marie-Louise von Franz, when you dream of an animal, the animal represents an instinctual drive that wants to be recognized. Its appearance is not necessarily good or evil; instincts carry both potentials and it is up to the ego and superego to guide them into conscious action. Dreams of shape-shifters, changing from animal to humans or blends of the two, mean that an instinctual quality wants to be expressed directly in a personal manner.

The wolf can be your noble and courageous ego force in action, or its dark side can be symbolic of greed. The bear may be warm, loving, and stable, or it may indicate a desire to be in total control of situations, moving a person to try to destroy anything that is not under his or her control. The raccoon may be clever and crafty or a thief. The cat may be warm and playful or a vicious, snarling beast. Among the Lakota tribe of the Great Plains, when people gathered at certain times of the year it was traditional to share such dreams, for they indicated special kinship bonds and consequent special abilities, resulting in the formation of clans of healers, much like the specialists we have today. Those who dreamed of bears had special skills with herbs. Wolf dreamers were good at treating people with arrow wounds. Dreams of buffalo meant abilities to quench blood flows. People who dreamed of elk were seen as charmers, as they have special love and sex powers—marriage and family counselors, and entertainers.

In typical snow goose fashion, in my dreams I was visited by snow geese as I was writing this chapter. They did not say anything or turn into people. They simply appeared and looked happy. Soon thereafter, a special opportunity came to go hunting geese and ducks for a weekend at the Spanish Springs Ranch in northeastern California. It was just before Thanksgiving, and around our home we try to have wild game on the table at all special holidays and birthdays.

The first day of the weekend was a goose hunt near Honey Lake, a great fertile marsh in the high desert along the Nevada–California border, which is one of the great waterfowl hunting places on the West Coast. We were sitting in a blind as the sun rose above the high desert, and with the day's first light a huge cloud of waterfowl rose above Honey Lake about three miles away. "Snow geese," said Harold, our guide. The birds circled the lake as more and more rose, creating a swirling vortex of snow geese that became a cloud drifting away to the east, toward the sun. There were several thousand birds, organizing themselves with perfect collective consciousness. We were mesmerized. "Won't come this way," Harold said matter-of-factly, "They go to a refuge and feed during the day and then come back to the lake at night." And he was right, they never came close.

Instead, our geese for the day were Canada geese, and out of twenty hunters, only two got geese, one of the lucky hunters being a ten-year-old boy on his first hunt. That night we moved thirty-five miles north in preparation for shooting on a chain of lakes at another ranch owned by Spanish Springs. In a dream that night I saw a Canada goose come to me, and then it was lying beside me.

The next morning we built a makeshift blind from sage brush on frozen mud beside a small opening in the ice on the lake. For the first hour there were ducks everywhere, and then things quieted down as the birds moved to a nearby refuge. From out of the still, icy morning air came the call of a Canada goose. I picked up my call and called back. The goose responded, and soon half a mile away a goose appeared over a ridge top and came flying in our direction. I kept up the call and the goose responded, heading directly for us. We were huddled together out in the open, trying to hide behind skimpy bushes. In the open water in front of us were six pintail decoys and no goose decoys, yet the goose kept coming. Two hundred yards away he set his wings and began to come in to land. Excitement was at fever pitch as we tried to stay still and get ready at the same time. A hundred yards away, the goose veered off, but still it kept calling. I called back. It went about two hundred yards away, then turned and came back, this time with wings set. It was coming in from the side, and we were totally exposed. I think that a flash of light reflecting off a gun barrel spooked it when it was just out of range, for it suddenly turned and headed off, honking loudly. The direction the goose chose, however, was straight over another blind at a mere thirty yards' elevation. As it reached them, it suddenly crumpled and fell, as if plucked out of the sky by a mysterious force that became known only when the sound of shots were heard two seconds later. It was the only goose shot that day. There was no good reason for that goose to be there. A mile away there were nearly a thousand Canada geese settled in a sanctuary lake. We had no geese decoys and were out in plain site. Did the bird want to come to us? I have seen ducks fly kamikaze-style into range for hunters standing out in the open, but never geese. They are very wary birds. If the willing-victim sacrifice of animals does exist, then this goose seemed to be living proof. The hunt was offered as a special one to introduce youngsters to hunting. One wonders if the Canada geese approved of what we were doing. It was the kind of experience with the spirit of the wild that makes a person understand why animals are sacred and feel great awe in their presence.

In the act of hunting, a man becomes, however
briefly, part of nature again. He returns to the
natural state, becomes one with the animal,
and is freed of the existential split: to be part
of nature and to transcend it by virtue of his
consciousness.

ERICH FROMM[9]

Two

The "Good Old Days"

We and our ancestors are the same people.
CARLETON COON[1]

An essential part of a waterfowl hunter's equipment are the decoys, representations that attract the birds to come within shooting range. You can buy good decoys through the mail or at a sporting goods store, but the ones that have the best spirit are handmade. Nearly all of these are made of wood, or at least they were before epoxy and plastic became available.

Wooden, floating decoys are made from two pieces of wood, one for the body, the other for the head. The wood for the body must be lightweight, easily carved, and durable. Aside from cork, the best body wood for decoys comes from aged cedar trees, which also make the best telephone poles. I learned to carve duck decoys from several older men, some of whom had been market hunters in the early 1900s. One of the first things they taught me was that when an ice storm hits you start looking for downed utility poles. New poles usually have too much creosote on them, so it's the old ones you want, because the creosote has been leached out by years of exposure to the elements. On days when one of those midwestern freezing rainstorms turns everything into a crystal wonderland, and tree limbs and utility poles are snapping like matchsticks, the only people you find on the roads are the repairmen and the duck hunters.

A decoy, in the final analysis, is art. If you were lucky and got to the downed poles first (after the crews had cleared the live wires), you would cut the poles into 14- to 18-inch sections with a chain saw and store them in the garage until they were dry. First you'd split them in half, then cut off as much extra wood as possible without destroying the basic form of the duck's body. Next, you'd draw the form on the sweet-smelling block of cedar, using patterns handed down from generation to generation with the exact proportions a carefully kept secret. After that, you cut out the rough form from the outline with a band saw. My high school shop teacher used to get quite a few bottles of liquor for Christmas every year for letting the duck hunters use the school's band saw to make decoy bodies.

The body is then shaped with a draw knife or "spoke shave," as the old-timers used to call it—a knife with two handles, one at each end, turned at right angles to a blade at least a foot long. If you're skilled with the draw knife, there is almost no need to use a wood file before final sanding of the body. Some people carve in grooves for wings and feathers, but you can't put in too much detail without running the risk of its being broken off the first time you drop the decoy into the bottom of the boat from icy hands that have just wrapped lines from a hundred other decoys as twilight fades and a stiff northwester blows at your back.

Decoy heads are more likely to be broken, so they should be made from harder wood for added strength. Pine or fir was always my choice. Leftover two-by-fours, if you know any carpenters, make the best heads. They would be thin enough to cut on a jigsaw. Again, head patterns were carefully guarded by the old-timers, in much the same fashion as shamans protected the magical formulas and dimensions of their amulets. A generic head really is an insult for the ducks. Canvasbacks have long triangular heads; bluebills and redheads have high foreheads; mallards, shovelers, teal, and widgeon have more flattened heads; and pintails have long necks with relatively tiny bills and large, roundish heads. The best eyes for all decoys are glass, countersunk into the head before painting.

Some modern folks use epoxy to glue on the heads. Old-timers used wooden dowels to link bodies and heads with extra strength, and brewed up a special potion from Weldwood wood glue that would make a garage stink for days to forever bond the heads to the bodies. To spend a day out in the marsh when the wind is howling and snow and rain are driving makes many people conclude that duck hunters have lost their heads, which may be so, but for a decoy to lose its head is unforgivable to a duck hunter.

Duck and goose decoys don't have to be perfect replicas of the birds to work. The one guideline that holds is that they can't shine. Flat paints are a must. In fact, lumps of dirt, rags, plastic milk jugs, and gallon paint buckets painted flat black have been known to draw ducks and geese into range. To be sure, some carvers can turn wooden blocks into forms that look more alive than the real thing, but it always seemed to me that the best decoys were the ones with the most character, even if they weren't the most accurate representations of the species they were supposed to attract. There was a look to those decoys the old-timers made, a crudeness that was more archetypal than lifelike, and it conveyed a spirit that seemed to draw ducks in like magnets, much better than any plastic molded blocks with all the feathers and fine lines. The Huichol of Mexico say that the hunter really hunts himself. Clearly, the old-timers' decoys had to have the same rugged qualities as the men who made them.

The oldest old-timers I knew were the last of the market hunters: people who killed ducks, geese, passenger pigeons, and anything else they could for money. It was a career that lasted only about two hundred years in the human history of North America, but during that time it was a life filled with natural magic that scarce few white people have ever known on this soil. Around the wood stove, while the winter snows drifted outside, these guys would spin yarns while they carved decoys. Tales, tall and true, would fly about like gusts of wind—the day we shot fifty canvasback drakes out on the cross dike; the time hundreds of black ducks got their feet frozen in a quick freeze and we went around picking up the birds like picking turnips in the garden; the day that flock of passenger pigeons flew over and we never saw the sun; or remember the time we got twenty-nine ducks out of one flock with just one gun. They remembered Prohibition, picking up barrels of whiskey the bootleggers threw overboard when the revenuers chased them up the river. They had lived a romantic life, not unlike the cowboy in American history.

These guys had known the "good old days," and they could serve up as many stories as there are ducks in the marshes of Lake Erie in early December. The hunter has always been a master storyteller, and these guys were among the best, but while their stories carried a lot of spirit, as the day came to an end, invariably they would sink into sadness about what hunting had become. Limits, licenses, plugs in your shotgun, game wardens, city hunters, and seasons had stepped between these men and the life they loved. They were no longer valued by society, and they had mostly their memories to live on. Many people blamed them for the extinction of some species and the declining populations of others.

The market hunter is generally pictured as an outlaw. To be sure, market hunting in the 1800s was responsible for the slaughter, even extinction, of some species. Birds like the great auk and the Labrador duck, which nested in large rookeries on islands, were especially vulnerable to seamen who gathered eggs. Shooting one-ton buffalo for their tongues, or just for the sheer sport of killing them and then leaving them to rot; dynamiting passenger-pigeon rookeries, killing young and old birds by the thousands as well as destroying the nesting trees; mass killing of egrets and herons in their nesting rookeries just for a few showy plumes for ladies' hats—these are a tragic page in American history. But stereotyping all market hunters as pathological butchers is hardly accurate or useful, except to flame hysterical opposition to all hunting and bring in money to animal rights groups. To blame market hunters solely for the extinction of the passenger pigeon, the heath hen, and the Carolina parakeet, and the near demise of the buffalo, is just not a full or accurate assessment of the profession of market hunting.

The wholesale killing, even slaughter, of wildlife by market hunters was permitted and even encouraged by the times in which they lived. They were not the outlaws of their times. Many were more like commercial fishermen today, without the benefit of good wildlife biologists to guide them. In recalling the extinction and near extinction of so many species in the 1800s, remember the equally important roles of habitat destruction, introduced diseases, introduced animals (rats, cats, dogs, mongooses, starlings, sparrows, and so on), and that in the case of the buffalo, government policy encouraging market hunting. Especially in the last half of the nineteenth century, market hunters were encouraged by the army to shoot as many buffalo as possible, to hamstring Indians by removing their primary source of food so that they would have to settle on reservations or starve. In *any* profession, if quick profits and fame are within easy reach, aggressive, greedy people driven by the will to power will come in to claim the bounty of power and money. There were no real controls over market hunting for too many years, and by the time people became aware of the damage that was being done, in some cases it was just too late.

Each species has its own niche and unique biological adaptations. Some are survivors, and others are not, as Darwin showed us. Survivors tend to be more adaptable, changing habitat and food requirements according to the times, and to have greater abilities to escape predation and protect themselves. Passenger pigeons, once the most abundant bird in North America and possibly the most common bird on earth, bred in massive colonies, were poor nest builders, and were not very intelligent. Imagine what it must have been like to walk into the breeding colony which settled into a 40-by-10-mile tract of woods near Petoskey, Michigan, in 1878. The pigeons flocked into trees in such density that large oak branches snapped under their weight. The females would lay one egg in a nest of just a few sticks piled together, and thirteen days after the chick hatched, the parents would leave the young to itself. The female might then move to a second nest and repeat the process with one more egg. That was it for the year. It has been estimated that each adult bird ate about a half-pint of nuts and seeds a day. One estimate of the flock's daily food intake is 17,424 million bushels.

As might be imagined, flightless chicks and injured adults fell to the ground under the roost trees in droves, where they were met by hordes of waiting dogs, raccoons, pigs, and foxes who gorged on the dead and wounded, while overhead the rest of the flock was oblivious. Passenger pigeons were ideal prey to greedy hunters, who used dynamite, nets, and even cut down trees to capture more than they could by shooting them with guns. One hunt of that massive Petoskey flock harvested an estimated 100 million birds. Adult passenger pigeons were the cheapest meat on the market in the mid 1800s.

Passenger pigeons were so vulnerable because they required massive tracts of forests for roosting, breeding, and feeding, and the success of their reproductive behavior depended on the over- all success of the colony, rather than on its individual members. Market hunters did overhunt them, mercilessly in some cases, but their extinction would not have occurred if they had been more like starlings or English sparrows in their breeding habits, and if the virgin forests of Michigan and other states had not been devas- tated by unmanaged, exploitive logging. A similar tale can be told for the extinct heath hen of New England, a chicken-sized par- tridge that became extinct in the 1930s due to the combined forces of habitat loss to farms, predation by introduced dogs, cats and rats, and overhunting.

Wildlife management was not a science until the 1930s, when Aldo Leopold became the first professor to specialize in the sub- ject. Little was known about the size of bird and mammal popula- tions or their migration patterns and habitat requirements. When market hunters came upon hordes of birds, mammals, or fish, they harvested as efficiently as possible, because that was their way of life. Profit drove them, and without regulation it sometimes turned to greed.

From the sparkle I saw in their eyes when the old-timers told me stories of the good old days, I could tell it had sometimes been a life of tremendous excitement. I do not condone exploitation, but I acknowledge that placed in a similar situation in that age, it would not be difficult to understand the attraction to market hunting. Among animals, weasels, foxes, and bears at times participate in killing frenzies similar to those of some market hunters. As a psychotherapist, I interviewed prisoners for the Seattle criminal justice system. None of the market hunters I knew fit in to what I consider a "criminal" personality. Criminals break laws knowingly, often flaunting authority as a game. They have low self-esteem and feel that they can't get what they want by trusting people or by working. Among those who go outside the law, some do so for survival needs, while others are in it for amusement, pleasure, or revenge. For the market hunters I knew, their occupation was the most enjoyable way of making a living. Some in fact had been heroes in their age. Their trade disappeared due to many factors that could not be tied together—habitat destruction, introduced diseases, lack of controls on harvesting, and the strength or fragility of each species' behavior and niche requirements—forces beyond their control and society's awareness in those times.

Most of the old-timers never seemed much at ease in a house. The four walls made them feel cooped up, so they tended to hang out in the garage. When the temperature dropped and drifts of snow piled up outside, they would stoke up the wood stove in the back of the garage, pull out a block of cedar, and carve. With a dog curled up on an old rug in the corner, they would swap stories about how it was when flocks of ducks that stretched from bank to bank moved down the Detroit River or the marshes came alive with spawning northern pike. Despite how good things might have been in their time, most of them fantasized about how it must have been when only the Indians lived there. Just the thought of the freedom of that lifestyle and their fantasies of abundance of game would cause a spark of life to jump through them that would double their progress in carving. The market hunters were a short-lived phenomenon in the history of hunting, an intermediary between the native subsistence hunters and the modern sport hunters. They entered a world of hunting motivated by subsistence and followed a new god of money to inspire their shots. Hunting was going on long before the market hunters. If Freud and William James were right in that hunting is instinctual in man, to better understand the hunter we will need to go back to the minds of those who hunted in the days before market hunting.

Native Hunters

Indians see storms as potential allies. Once I set up a meeting between Rolling Thunder, a noted medicine man, and some members of the Northwest Coast tribes. When you meet a medicine person it is customary to give him or her some tobacco or an herb, like angelica. On this occasion, as we were waiting to see Rolling Thunder, a man asked me if I would like to see the gift he had brought for Rolling Thunder. I said yes. He reached into a buckskin bag and pulled out a jagged piece of wood. I looked at him questioningly. "Lightning struck this tree," the man explained, putting the chunk back into the bag. "Wood that comes from trees felled by lightning has special powers. It carries with it the spirit of the thunder people, the spirit beings who live in the west and shoot lightning bolts of truth into people's hearts." The wood was a very appreciated gift.

Out along Puget Sound, Indians wait for winter storms to topple giant cedar trees. Among tribes like the Lummi, who live in northwestern Washington near the Canadian border, old-growth cedars are not just big old trees—they are wise elders who can exude wisdom to the field of collective intuition and even talk in your dreams. Old-growth trees are sacred, the Lummi say. Stands of old growth are their cathedrals. According to tradition, you have to ask permission of the tree to take its bark or cut it down;

show proper respect by saying prayers and leaving a little food or tobacco as a sacrifice before taking any wood, even if the tree has fallen in a storm. When the old growth goes, the spirit of the land weakens and life becomes harder, Northwest Coast Indians say. One of the most important elements of native culture is that there is a spiritual landscape everywhere, as well as a physical one.

In this land of giant trees, duck and goose decoys can be carved from wood or you can take a bunch of twigs from the cedar trees and bind them together with bark so that they look like real ducks. If you want to get fancy you can also skin the ducks you catch and tie the skins onto the bundles of cedar twigs to make the best decoys of all.

A major difference between the native hunter and the modern hunter is that the native begins his efforts to attract game long before he goes into the field. Art in Paleolithic caves, such as Lascaux in France or in the outback of Australia, frequently depicts wild animals and hunters pursuing them. We can only guess on the original reasons why these pictographs were made, but invoking sympathetic hunting magic is a very common motive.

Before mechanical technologies empowered humans, magic was a principal tool for increasing personal power. When the hunter crawls into an underground cave, fasts, foregoes sex, makes offerings, plays drums or shakes rattles, or builds a fire and performs songs and dances that have come to him or her in dreams, the hope is that a sympathetic chord will be struck with the animals, their spirits, or the spirits of their guardians, and this will increase the chances for a successful and safe hunt. For the traditional hunter, the paintbrush and the rattle are the forerunners of the spear and the bow and arrow. A dance, a ritual, and a story are the first attempts at calling game within range.

Sometimes the old-timers would joke about how you have to think the right thoughts to get close to game, especially the wily animals. It was one of those jokes that everyone likes to laugh about, but I'm not so sure that in private their attitudes were so different from those of their Paleolithic predecessors. Only the hunter knows what goes on in his or her mind when the hunt is in process. Hunting psychology, then and now, holds many closely guarded secrets. Focused intention is necessary for mastery in any activity.

In many kinds of hunting, a hunter uses calls to attract animals. Some replicate the sounds made by animals, such as duck, goose, turkey, quail, and deer calls. Others produce sounds that resemble food, such as the wounded rabbit call that attracts predators including coyotes, wolves, bobcats, and birds of prey. Old-timers can tell the brand of a call a quarter of a mile away, as well as who is using it. Newcomers always put the wrong emotion into a call, because they're too excited, the older hunters taught me. The only way you can really become an expert is to learn to think like what you're trying to attract. Calls, as you might expect, are coveted musical instruments as well as tools of ritual magic. There is an equivalent of the Stradivarius for calls, and believe me, if you own one, it is among your most prized possessions. The wood, the vibrating reed, the shape of the call, its pitch, how hard you must blow to activate it, and how softly you can blow without losing the proper tone all play key roles in making a hunter successful at calling game.

Good art is a voice of the land. The most famous art of the Northwest Coast tribes are totem poles—large trunks of cedar trees that are carved into a series of animal forms, some real and some mythic, that are stacked on top of each other. Not many people carve totem poles these days. Some of the best are preserved in museums such as those at the University of British Columbia and the University of Washington. In their presence one feels the meaning of spirit.

In traditional villages such as Skidegate in British Columbia, outside of each family's longhouse was a totem pole. Ravens, eagles, beavers, seals, otters, salmon, killer whales, and bears are among some of the animals most commonly carved into their designs. Modern sculptors use their art to make a statement, expressing their views and projecting their unconscious into form. Some modern art has appeal because for most of us, the unconscious is a wilderness yet to be explored. The real power of the artist is to make us aware of our souls. The native arts generally are not so much statements about feelings as a reflection of kinship with nature, and, perhaps, an invocation of the spirits of our relations. Traditional arts may be trophies, too, proclaiming that so-and-so has seen such and such a totem animal in a vision. Boasting is an eternal pastime among hunters.

Today we walk through museums and study the totem poles, admiring the carving techniques and evocative forms. In its native land, a totem pole is not the equivalent of a statue, like the religious icons we place in houses of worship. These poles are like the Scottish clan shields or family crests of the British Isles that contain animals such as deer, elk, boar, badger, and fox. Genealogy among native peoples extends into the world of plants and animals. Totem poles proclaim special kin lineages in the natural world and identify each family as having the special powers associated with it. We all are descended from people who believed they were descended from animals, but, having become modern, we just joke about our animal nature. We all know people who are sharks, they become attorneys. Bears and rams become politicians. Wolves become musicians and writers. There are human rats, too. The animal kingdom is still very much alive in modern language, as well as in the modern psyche.

Nara, Japan, is the site of the origin of the Tenrikyo religion, a modern international offshoot of Shintoism with more than 4 million members. The religion began at 8:00 A.M. on October 26, 1838, when a forty-one-year-old Miki Nakayama began to receive divine messages directly from God. Today, on the exact spot Mrs. Nakayama received her first calling, stands the largest wooden building in the world, a striking complex the size of four football fields. Precisely in the center of this temple, a simple wooden obelisk—the "jiba"—rises from the floor, marking the exact location where the inspiration took place. Twice daily, pilgrims by the thousands come to pray in this shrine, encircling the sacred jiba, performing chants, prayers, and hand claps in unison. Nearby is an impressive collection of canes and crutches that people have discarded as a result of miraculous healings experienced during prayers. This is a place of power. You can feel it in the air as you approach the temple. The numinous is everywhere here.

At special times, the *Karuga* ceremony is performed around the jiba to honor creation. According to the teachings of the founder, human beings started out in the oceans, and only after nine hundred million and ninety thousand years did they take form on land. Before that time, men lived as mermen, which evolved into human males by combining with the killer whale. Women began as turtles, which then transformed into human females when they combined with the white snake. Carrying these ancestors with them, humans were created when the first male and female were born from the she-monkey, who is the mother of us all.

Your family tree is not just a set of lines of heritage; for most people there is also a special species of tree, flower, or shrub associated with it. I'm a member of the Gunn clan, the northernmost clan of Scotland, and my clan shield has a sprig of juniper on it. Other Scottish clan shields carry oak leaves, holly, and mistletoe, the Druids' wonder plant.

Like American Indians, Scottish chieftains sometimes wear eagle feathers on their heads and carry belt pouches made of badger and bear skin. Many clans and families include animals as part of the symbol or name that unifies the group and sets it apart from all others. The origins of these associations with certain animals remains cloudy. It is obvious why a chief might want to have the keen eyes of an eagle or a warrior might want to have the courage and strength of a bear. The Earl of Northumberland claimed that his mother had mated with a bear and so a bear appears on his family crest.

Australian aborigines look for special, familiar links between people and animals from the moment of birth to tell just what kind of personal power someone might have. Other tribes might undergo rituals to seek visions and dreams in which totem animals would appear. This link between man and animal was once considered by some Christians to be the mark of the devil; the witches' familiars included goats, dogs, cats, toads, and owls. This is where we went wrong and fell from grace; equating nature kinship with the devil is really a projection of the fear of one's instinctual nature, which if allowed to come forth would violate Victorian standards. In the watery depths of our unconscious, the symbols of animals and plants still live. Everything in the psyche is alive, Carl Jung pointed out, for it is a part of us.

My good friend Kenny Cooper, Lummi Indian Seowyn spirit dancer, tells me that among the Salish-speaking tribes of Puget Sound there is a word that we should all learn: *skalalitude*. There are only about three thousand Lummis. They are one of those rare tribes still living on their ancestral lands, and so they have a peacefulness about them, like the Hopi in Arizona or the Winnebago in Wisconsin, that you don't often find on Indian reservations elsewhere. Skalalitude means that when we are in right relationship with nature, there is magic and beauty everywhere. There is no similar word in the English language, which indicates our uneasiness at the idea of mental unity with nature, as well as our forgetting of just what value nature kinship may have in human life.

Kenny's Indian name is Cha-Das-Ka-Dum. It means "man in the moon." But Kenny looks more like a grizzly bear or a Sasquatch than the moon. He's about six-foot-three and tips the scales at the Lummi fishing dock at close to three hundred pounds. His traditional costume is a black bearskin suit with many small hand-carved canoe paddles circling his barrel chest. On his head is a cone-shaped wizard's hat made of cedar bark. At his belt is an otter-skin medicine bag for carrying special ceremonial objects. The otter is a playful creature that enjoys life, and Kenny is at his best as an entertainer with stories, songs, drum, and flute. In his traditions, wardrobe design comes from the spirit world. All these items came to him in dreams and as a result of traditional teachings. The Lummi, along with most other native tribes, believe that when you treat nature with respect, there will be reciprocity.

The glacier-fed Nooksack River runs through the Lummi reservation, which is just north of Bellingham, Washington. I remember the Nooksack well because when I was teaching at Western Washington State University in the early 1970s, I loaned my canoe to two students, and they proceeded to swamp it in front of a log jam on the Nooksack. The canoe was never seen again. I guess it was my sacrifice to the Salmon woman, whose spirit oversees the salmon runs up the rivers of Puget Sound every fall.

The Lummi have a custom of purifying their minds and bodies by going up into the mountains near snow-capped Mount Baker, seeking out those icy cold glacier-fed streams that lead into the Nooksack River and jumping into special deep holes in the streams. Then they pray and fast, hoping for visions, dreams, and voices. Traditionally, all Lummi kids are brought up to listen with the "third ear," the heart. Kenny says that if you want to see his connection with the moon, just look up there and you'll see the right-side profile of an Indian kneeling. When Kenny kneels down to pray in this position, I can see the similarity.

The moon carries the power of dreams, water, and intuition and holds influence over the weather and the creatures of earth and sea. Kenny advises that before you go hunting, make a prayer and give a little food or tobacco to the spirits of the wild. That way you'll show proper respect, increase your chances of seeing what you want to catch, and avoid accidents. In the Lummi view, nature is a whole, a storehouse of food to the fortunate, and a source of wisdom to the blessed. When you hunt in the traditional way, you enter into the web of consciousness of the creatures and their guardian spirits and become one of them. The traditional hunter goes into the field in search of food, just like his modern counterpart, but his mental attitude is different if he is following the pure spirit of the hunt.

One of the most-quoted modern guides to hunting is José Ortega y Gasset's *Meditations on Hunting*. The Spanish poet's work remains a classic in hunting psychology, for modern people anyway. Ortega's prose captures the magical feelings of stalking game, when man and animal blend into one web of consciousness, yet he also says, "Every good hunter is uneasy in the depths of his conscience when faced with the death he is about to inflict on the enchanting animal."[2]

Ortega may speak to the psychological dilemma of modern man, who has grown up in a cozy home far from the cries of geese flying across the face of the moon and the bugling of a bull elk in rut. Guilt judges an action, declaring it wrong with a twist of the knife in the stomach. Guilt is usually the emotion that fuels the burning ulcer in the stomach of the workaholic executive. Guilt is the most significant emotion that man must face and resolve if he is to become a successful hunter. It is part of growing up to learn what one should feel guilty about. Among animals that hunt, guilt does not seem to exist. Ask any cat or his big cousin, the cougar, if he feels guilt when he has killed another animal, whether mouse or deer. He'll lick his lips.

In general, the closer we are to someone or something, the more difficult it is for us to kill it, for we are more likely to feel guilt for committing that act. For most people, it is much harder to kill a person than an animal. In his memorable line from the Academy Award-winning film *Unforgiven*, Clint Eastwood as William Munney says "It's a helluva thing to kill a man," and we will never forget it. In war, soldiers are trained to kill the enemy by propaganda that pictures them as inhuman and evil, to lessen guilt feelings that can dull the competitive edge in combat. High-tech wars are fought with computers, creating more psychological distance between the killer and the killed, but the deaths are the same. The native hunter cannot escape the feelings of the hunt. The hunt fills his or her entire life. It is more difficult for the modern hunter to understand as much about the psychological issues of hunting because it has become an activity to be crowded into one's schedule, rather than being the core of one's life and identity. We have become more like the dog than its ancestor the wolf.

From studies of modern psychopathology, we know that only the psychopath seems unconcerned about guilt when he kills. Once you have felt guilt knot up your stomach, you know you don't want to feel that way again. Guilt protects the innocent. However, the traditional hunter, who is very close to the animals he hunts, tends not to be troubled by guilt when he kills animals. His concern is more with fear. The Blackfoot of Montana tell the story of a famous hunter who had buffalo medicine. His gift endowed him with special skill in driving large numbers of buffalo into a corral called a *piskin*, where they could easily be slaughtered. One day, the story goes, he woke up and ran out and burned down the piskin. The tribe was furious and demanded to know why he had done it. He explained that he'd had a dream where the buffalo spirit had come to him and told him that he had killed too many buffalo because of his special kinship with them. The buffalo spirit in his dream had told him that to show that he had not lost respect for the spirit of the animals he hunted, he should destroy the piskin and pray for forgiveness, otherwise bad luck, food shortage, and sickness would befall the entire tribe. Upon hearing this story, the tribe understood and joined his prayers.

The traditional hunter used whatever tools and techniques he could to catch food: rocks, clubs, spears, pitfalls, snares, dead falls, driving animals over a cliff, and knives. They say that around San Francisco Bay, Miwok and Pomo tribes once hunted ducks by swimming up underwater and grabbing unsuspecting ducks by their feet; a method that snapping turtles still use. The forerunner of the roadkill was the mastodon trapped in a swamp, the bison herd that stampeded during a thunderstorm and plunged over a cliff, and the migrating ducks that were exhausted and landed in a pond that froze overnight. All such "accidents" were gladly accepted, along with anything else that could be caught. Subsistence hunters enjoy the chase, but their bottom line is food. Regardless of his technique, the traditional hunter prays that he may kill quickly and cleanly, and he shows deep respect for the animals killed. His feelings of guilt are lessened by several factors, one of the most basic of which is a very pragmatic one; hunting is necessary for survival.

Anyone who takes seriously the need for modern people to better understand aggression and violence should read Erich

Fromm's benchmark book *The Anatomy of Human Destructiveness*. In this extraordinary look at the dark side of human nature, Fromm reviews the shadow from as many different angles as possible. I will let him have the last word on the motives and nature of native hunters:

> Our knowledge of hunting behavior is not restricted to speculations; there is a considerable body of information about still existing primitive hunters and food gatherers to demonstrate that hunting is not conducive to destructiveness and cruelty, and that primitive hunters are relatively unaggressive when compared to their civilized brothers.[3]

Modern man's food is a poor substitute for the original health food, wild game. Seventy-five percent of the "diseases of civilization"—heart attacks, diabetes, strokes, kidney ailments, atherosclerosis, emphysema, hypertension, cirrhosis—were rare or nonexistent in the Paleolithic era, when Kenny Cooper's ancestors were out on the Olympic Peninsula chasing mastodons with spears, perhaps hoping to drive one over a cliff or into a special pit, according to S. Boyd Eaton, M.D., Marjorie Shostak, and Melvin Konner, M.D., in their book *The Paleolithic Prescription*. Eminent anthropologist Ashley Montagu calls this book ". . . probably the best and most useful diet, exercise, and design for living book ever written."[4] The researchers report that the serum cholesterol levels of hunters and gatherers and other tribal peoples are considerably lower than those of Americans and Europeans, even though the natives may have more meat in their diet. But as soon the native peoples take on a modern diet and lifestyle, their cholesterol levels immediately rise.

Before the television set replaced the fire pit, the refrigerator replaced sun drying, and trucks replaced canoes, the major causes of death were infections, injuries, trauma, childbirth, starvation, and weather (exposure). In general, wild game has one-seventh the fat of beef, more protein and minerals, and less cholesterol. Nutritional research among the Northern Cheyenne elders shows that if the old people could return to a low fat, high protein diet of wild game meat, the incidence of heart disease, kidney disease, and diabetes would drop significantly. The circumpolar tribes, such as the Eskimo, when they exist on an almost all-meat diet, live to ripe old ages without the diseases of our civilization. A key here is that they eat the inner organs of the animals, not just the choicest steaks, for organs contain vitamins and minerals crucial to a balanced diet.

Among the Inuit, or Eskimo, like many other tribes of hunters and gathers, weapons were believed to have a special spirit that helped the hunter make quick kills. Today, the traditional hunter who hunts with aboriginal technologies such as the boomerang, bow and arrow, sling, snares, and dead falls is extremely rare. Given the money to buy one and keep cartridges in stock, the modern aboriginal hunter will prefer to use a rifle or shotgun, an outboard motor, and a snowmobile. There are perhaps a 250,000 hunter-gatherers remaining today, living as their ancestors did 10,000 years ago: pygmies, !Kung Bushmen, hill tribes of India and Ceylon, Andaman Islanders, Australian aborigines, Ainu of northern Japan, and some of the Inuit and other tribes above the Arctic Circle. To survive without modern weapons or other tools, the truly traditional hunter must cultivate faculties and senses we moderns have little or no familiarity with. He cannot afford to go home empty-handed very often, otherwise he and his family will die. The idea that the traditional hunter hunts to torture animals for sadistic pleasure, as advanced by some animal rights activists, is ridiculous. Hunter-gatherers have no supermarket to forage in if the quarry gets away. Tortured animals escape when they can, and there is no certainty of when the next one will be caught. The prayer for a quick, clean kill is shared by all true hunters in human history. And for the traditional hunter, the idea of torturing an animal would be unthinkable because the spirits of the animals would retaliate.

Because of the basic need to kill in order to live, to balance his lack of technology, the traditional hunter chooses to perfect human skills with which we have but the slightest familiarity. What modern psychology fails to understand is that much of human perception is developed according to choice and cultural conditioning. To recognize tracks and signs of the trail of a deer requires an eye specially trained to notice telltale anomalies: the slight imprint in the mud, the crushed leaf, the bent twig. Studies of native peoples show that they have developed a much keener sense of hearing than we who hunt in supermarkets bother to cultivate. A Chippewa friend once told me that his equivalent of a master's degree in hunting was earned the year he spent learning to track game with his sense of smell. During that time he was blindfolded and had to get down on all fours to sniff the track, just as a dog would. We can become like the Northwest Coast Indian paintings that show bodies covered with eyes. Such awareness is a skill, just like shooting a three-point basket, tasting wines for their bouquet, or giving a healing massage.

Having cultivated his sensory awareness, the traditional hunter is skilled at tracking and stalking. He is like the wolf or the panther. The traditional hunter balances the limitations of his killing technology by entering into the mental realm of the animals and seeking to outwit them; in Samoa they have a saying that "the fish do as the will of the master fisherman." To enhance this mind-set, ceremonies are performed and other special customs are kept before, during, and after the hunt. There are entire cultures that are based on songs, stories, myths, legends, dances, and ceremonies that promote sympathy and respect between man and animal, and sympathy is a key element of intuition and telepathy as well as magic. In sports psychology, the mind-set of optimum performance is called the "zone." Traditional hunters were the pioneers of this zone state of mind.

Officially, modern psychology is extremely skeptical of intuition, let alone telepathy, clairvoyance, clairaudience, and clairsentience. It calls such extrasensory states paranormal or, more often, foolish superstition. Modern psychology admits to only five senses with which we perceive the world around us. Native American shaman Rolling Thunder asserts that what is paranormal to modern people is normal to Indians. The ancient wisdom of China asserts there are at least one hundred senses that can be used to perceive the world around us. They include abilities such as cultivating the art of dreaming, the ability to sense and feel the electromagnetic and other subtle energy fields of the earth, the ability to use one's entire body to sense the ever-changing pulses of gravity, the periods of the moon and the tides, the ability to recognize the directions of the compass, and so on, as well as the "psychic" abilities. For the traditional hunter, the entire body and mind are one giant sensory organ. "We now touch upon a problem of the greatest importance . . . that is, the question of the *reality* of the extrasensory capabilities ascribed to the shamans and medicine men. Although research into this question is still at the beginning, a fairly large number of ethnographic documents have already put the authenticity of such phenomena beyond doubt," concludes Mircea Eliade, considered by many the most prominent religious historian of the twentieth century.[5]

In support of Eliade's claims is A. Foster's study of Indian children in the Manitoba public schools. Psychic abilities are hard to produce on command, because such testing defies the holistic nature of intuitive perception, but nonetheless, when fifty Indian children were shown standard ESP cards, their accuracy in recognizing the symbols on the unseen cards was far above chance.[6] In American Samoa today, all schoolchildren are given a chance to test their psychic skills with the ESP cards used in tests of telepathy and clairvoyance. This unit has been included in the curriculum at the request of the Samoans to ensure that the traditional Samoan mind is respected when intelligence is calculated.

When Kenny Cooper wants to go hunting out on the Nooksack flats, just north of his home, he says his prayers and leaves offerings to a world that is all-knowing and alive. Some tribes feel that this honors the spirits of the animals. Others believe that a mythical being or god oversees nature and animal life. Among the Eskimo, the keeper of the animals is Sedna, or Pinga, the woman who lives at the bottom of the sea. In West Africa, the offering would be made to the god Ogun, who watches over the work of hunters and the blacksmiths who make their weapons. Regardless of whom one honors, the practice of making sacrifices before going hunting is very common among native cultures around the world.

When Kenny was growing up he learned to hunt deer with wooden arrows and a bow made from the yew tree, but today he uses a gun and does quite well getting his deer for the winter. He no longer dons buckskin and uses a bow, but he maintains that the spirits are still there, helping bring him together with the right animals to shoot. Because modern technology is so potent, Cooper says, we should become very religious about hunting, as well as life in general, so as not to make mistakes. With power comes responsibility. Balance is required.

The most significant difference between the traditional hunter and the modern hunter is that the traditional hunter lives in a world in which "paranormal" forces play a major role in all aspects of life, including the hunt. These parasenses play a more important role in his choice of orienting faculties, and their use is supported by the culture within which he lives. As we move into a more multicultural society, modern science is going to have to loosen its denial of extrasensory perception, or charges will begin to be leveled that its paradigm is racially discriminatory. Modern man is uneasy at the thought of spirits, except perhaps for the holy spirit or the holy ghost of the Christian trinity—father, son, and holy spirit. To say that you believe in spirits of the land, the animals, the stones, the plants, the world above where the angels live, or the world below where the spirits of the dead live, is to be a follower of animism and perhaps a pagan. It can also be seen as a symptom of psychosis. Today, fiction is perhaps the most honest educational medium we have to teach about the experience of life. Remember in the film *Field of Dreams* how the farmers in the feed store reacted when Kevin Costner said he'd heard voices in his cornfield? No one categorically dismissed him.

We moderns hunt "spirits" in taverns. The traditional hunter offers nature spirits food, tobacco, or herbs such as angelica root, which is much prized by my neighbors the Pomo and Miwok tribes of California. Any hunter who has spent time alone in the wilderness, far from the confines of clocks, schedules, walls, and windows, knows that forces beyond our control are behind the meeting of hunter and the hunted. In the four hundred generations that have passed since the days when the Wisconsin glacier receded, some ten thousand years ago, there has not been enough time for notable genetic changes. As Robert Ardrey has said:

> For millions of years we survived as hunters. In the few short millennia since our divorce from that necessity there has been no time for significant biological change—anatomical, physiological or behavioral. Today we have small hope of comprehending ourselves and our world unless we understand that man still, in his inmost being, is a hunter.[7]

Regardless of whether you believe there are discarnate entities that exist in realms beyond the temporal, and that they can and do influence life in this dimension, perhaps even being a major force in the cause of events, health, and disease, try to suspend judgment for a moment and see the world as it exists for native peoples. This will help you understand what the "good old days" were really like. Among hunters there is a great mystique about the "native guide." We need to comprehend what makes him or her so special if we are to better understand the mind and soul of the hunter. To do so, let's look at examples of how three different animals that are hunted—the goldeneye duck, the deer, and the bear—are viewed in those cultures where subsistence hunting still is practiced.

A Totem Pole of Meaning

1. The Goldeneye Duck

On Lake Erie, as the duck season wound up in December, one of the last flights to come down from the north was the goldeneye. The large black-and-white males with a white dot under their eyes are a striking sight. "Whistlers," we called them, because their wing beats make a loud whistling sound that, on a calm day, can be heard several hundred yards away. We seldom shot whistlers. They tend to eat fish, not grain, which gives their flesh a strong taste. But I have always felt a special enchantment when watching whistlers speed by, their musical flight sounding like bells on the cold north wind, messengers of the spirit of winter.

We each see the world according to our heritage, as well as any commonly shared senses. In my family tree, the root that come from Scotland traces back to the Saami, or Lapps, of Northern Scandinavia, the people who gave us many of the mythic images of Christmas, including those of a sleigh pulled by magical flying reindeer and of a man in a red suit who lives at the North Pole, gives presents to good children, and likes to have people celebrate around a pine tree. Perhaps my fascination for the goldeneye is due to my Saami ancestry, for in their legends the goldeneye is a bird of great mythic importance.

The goldeneye plays a principal role in the creation of this world, according to the Finnish epic poem the *Kalevala*. The world began, the *Kalevala* says, when the Air Spirit grew restless and came down to earth. She settled into the sea and soon a gust of wind blew her pregnant. As her pregnancy progressed, she felt lonely and prayed for companionship. Soon, a goldeneye duck was sent from Ukko, god on high. The tiny duck flew to the four directions and finally landed on the knee of the Great Mother. There the goldeneye built a nest and laid six golden eggs and an iron egg. The maiden and the duck brooded the eggs and in time they became very hot. The Mother of all could not stand the heat so she shifted her knee, sending the eggs into the icy waters, where they broke into pieces. One became the sun, while others became the moon, the sky, the stars, the clouds, and still another became the earth.

Every wild creature is special, and each calls out to someone with a unique harmony that reaches down into the very deepest recesses of the mind. But if you are descended from the Saami, the sight and sounds of the goldeneye reminds you of your origins. The duck becomes a carrier of myths and a preserver of legends.

It is not hard to see how a bird like the goldeneye could become a mythic symbol, and all around the world creation myths of native peoples tell of ancient times when animals could talk, gods walked the earth in physical form, and the two got together to create life as we now know it.

While the animals vary from land to land, among all native peoples I know, special species of animals are considered to have unusual mythic and spiritual importance, which makes them central figures in the local religion. These animals live both in the world around us and inside us in our unconscious mind. Within each of us there is a totem pole that gives meaning, power, and purpose to our lives. Looking briefly at two common animals, the deer and the bear, illustrates how whole systems of cultural sentiments can be built around spiritual reverence for animals that are hunted.

2. The Magical Deer

Deer are the most popular animal that draws hunters from their beds out into cold weather (except in California, where the deer season is in the summer). For deer hunters, tiny towns in northern Michigan such as Oscoda, Grayling, Alpena, Drummond Island, Hale, Manistique, Kalkaska, Newaygo, and Crystal Falls are like the pilgrimage places along the route to Mecca. Each November, nearly a million men and women dressed in red and orange suddenly make a pilgrimage to these places of power. They pump more than $300 million into the local economy, and their license fees contribute more than $14 million to support the Michigan Department of Natural Resources.

There are similar deer pilgrimage places in Pennsylvania, Wisconsin, Minnesota, Texas, Wyoming, Montana, and other states. More than 3 million deer are taken by legal hunters every year, putting some 18 million pounds of venison on the table. To be sure, when the deer season is open bars swell at night, filled with ruddy-faced hunters boasting or complaining, but there is something else going on, too. Deer camps are not just underused vacation cabins. With antlers, animal skins, old photos, and historic logging equipment hanging on the walls, there is an air of sanctity about hunting cabins that you will not find in most other dwellings of modern man. These cabins are the modern equivalent of the temples of secret hunting societies that have bonded hunters together for thousands of years. More hunters feel this way than are willing to admit it publicly.

Today, in Michigan, the deer population has grown dramatically. Before 1800, most of northern Michigan was covered with a dense pine forest and deer were scarce. Only in the south, where there were plentiful acorns and lots of swamps, were deer common. Altogether, maybe half a million deer lived in the Wolverine State in the days before logging.

As settlers moved in and cleared land for farms, deer numbers increased as habitat increased. Then came the market hunters, and by 1885 there were almost no deer in southern Michigan. Meanwhile, in northern Michigan, as logging leveled the white-pine forests, deer flourished on the grasses and shrubs that sprang up. By 1870 the Michigan deer population had doubled to a million in the northern part of the state.

As the herds mushroomed, the market hunters followed. Deer were trapped, snared, and shot by the hundreds of thousands. In summer, when the meat could not be stored, killing continued for the skins, leaving the woods dotted with dead deer in a fashion similar to what was being done to the buffalo herds to the west.

Along with the market hunters came hardwood logging, and the glades and small openings in the forest that offered food and nearby shelter began to disappear. The landscape now became forests of giant stumps, ghosts of the virgin white-pine forest. In summer, fires raged through the piles of branches and smaller trees that had been cut to remove the giant white pines. By 1900 there were only 50,000 deer left in Michigan; nationwide, there were perhaps 500,000 whitetail.

Forest fire control now entered the picture, and with it the barren lands began to grow again, this time filled with aspen, cedar, and other prime deer foods. The deer population skyrocketed, and by 1925 there were numerous examples of overbrowsing. Biologists believe that by 1950, 50 percent or more of the fawn crop of that year died of starvation in the cedar swamp winter yards, where whitetails go for shelter during the heavy snowfalls. In the late 1950s, I vividly recall walking through the winter yards in the swamps where the lower branches of the white cedar were pruned off like a British gardener had been trimming hedges at four feet. Above that level was a dense, green canopy of deer food, but it was all out of reach. What I remember most vividly was the stench: rotting deer carcasses and skeletons were everywhere.

From 1950 to 1972 the Michigan deer herd plummeted from 1.5 million to 0.5 million. The primary causes were, first, maturation of the forests (the leaves and twigs of the second-growth trees grew out of reach and also shaded the shrubs, grasses, and other browse on the forest floor), and, second, the overabundance of deer had eaten everything in sight. They began to starve in massive numbers.

The state legislature took action in 1972, passing a law that allocated $1.50 from each deer-hunting license for deer-habitat improvement. Over the next twenty years, approximately $20 million was spent on cutting forest openings and buying up wintering yards. Today, with habitat restoration and careful monitoring of deer herds, the Michigan herd has grown to nearly 2 million. A healthy doe gives birth to twins or triplets every year, so each spring some 900,000 fawns are born. By October 1, when archery season begins, 500,000 of those fawns will still be alive. During the three seasons—archery, rifle, and muzzleloader—that last from early October to late December, hunters in Michigan kill more than 400,000 deer a year, nearly a 50 percent success rate, the second largest harvest in the United States, and even that isn't enough to stabilize the herd population. Despite the size of the harvest, deer kills by car have jumped, crop predation has increased, and as many as 125,000 deer a year starve to death. Managing the Michigan deer herd these days is even more challenging; as 75 percent of the deer now are found on private land, an increasing amount of which prohibits hunting.

Michigan is liberal in issuing permits to hunt antlerless deer, which is a preferred management strategy by wildlife biologists. In 1991, 267,479 permits to hunt does or young bucks were sold. One consequence is that 25 to 35 percent of Michigan's fall herd are mature bucks. Contrast the Michigan herd with California's. There are close to a million deer in California, of three species—blacktail, mule and whitetail—and the carrying capacity (the maximum number the habitat can support) of most areas is at the limit. There were more deer in California in the past, but between 1950 and 1980, 4.8 million acres of wilderness were consumed by development. In 1991, the 225,000 California deer hunters harvested 44,500 deer, nearly all bucks, an 18 percent success ratio. The kill by illegal poachers may be as much as twice that number.

Very few permits to hunt antlerless deer are given out in California, due to political pressure, not to the advice of the biologists. One result is that only 10 to 15 percent of the fall population are bucks. A second consequence is that the annual fawn crop is almost as large as Michigan's—700,000. What is shocking is that according to Department of Fish and Game research, 75 percent of these fawns will die of natural causes before they reach their first birthday. A prime cause of early deaths among fawns is starvation, because the abundant adult does have already consumed much of the feed that the fawns could reach. Most of California has a Mediterranean climate, with wet winters and hot, dry summers. Hence, fawns born in the spring must make it through the summer heat and drought before the abundant foliage and grass returns in late fall. This is why my wife's purple petunias on our patio are so popular. Anyone who says that hunting causes excessive cruelty and suffering has to look at what happens here when deer are hunted so selectively according to gender. The average time to death from the moment an arrow or a rifle bullet hits a deer in the heart-lungs kill zone until its heart stops beating is less than thirty seconds. Starvation and disease are long, slow deaths with considerable suffering. A third consequence of the restrictions on harvesting deer in California is that the state's Fish and Game Department has a lot less money to work with, which is one reason why poaching is so high. In 1992 there were only 293 game wardens in the field in California, which studies conclude is a major reason why poachers have a 99.5 percent chance of not getting caught.

Today in Michigan, as in other states, farmers sell "deer food" to hunters for their deer camps, making alfalfa, sugar beets, and apples an important cash crop. The story of Michigan deer is repeated throughout the Midwest, the South, and the East—wherever the adaptive whitetail lives. Conservative estimates place the herd in the United States and Canada today at 27 million. Pennsylvania reports more than 30,000 deer-car collisions a year. The automobile has taken the place of the bear and the panther as the major predator of deer. Nationwide in 1992, 460,000 deer were killed by autos, and 159 people died in these accidents.

Deer are survivors: hoofed rabbits with antlers. Man has always been the most significant predator of deer. The size of the white-tailed herd in many states is currently a hot political issue. My prediction is that the issue will increasingly appear on ballots as animal rights advocates try to ban hunting and as deer learn to enjoy the pleasures of suburban life.

Deer have a numinous quality. Chinese folk wisdom believes that the presence of deer in an area is a good omen, which agrees with American Indian beliefs. Deer are also feisty. As many as 10 percent of all bucks are killed in breeding battles each fall. In

North America, because of their sheer numbers there are more confrontations between people and deer than any other wild animal. Bambi of either sex can be dangerous.

Among the Huichol of Mexico, to whom the deer is the symbol of their tribal identity, often the deer is portrayed in art as having two faces. One side is old and wise. The other has a devilish grin, for the dark side of the deer is lust. According to Karuk shaman Medicine Grizzly Bear, if you see a deer on a vision quest it may mean that you are going to encounter a seductive person of the opposite sex in the near future.

Deer grow a new set of antlers every year, and the antlers have a special magical power for many people. Shamans around the world are pictured wearing hats with deer antlers. Chinese geomancers say that plentiful deer in an area is an omen of good chi in that place. Traditional Asian healers brew up an aphrodisiac potion from deer antlers. The best known mythic animal in modern times is Bambi, closely followed by his cousin Rudolph. Saint Norbert and Saint Hubert both had visions of Christ appearing over the head of a stag. Analytically, one might interpret such a vision as a calling to raise one's lustful physical energies to a more spiritual approach to life, in which one focuses on serving humanity's needs rather than one's own. A simplistic approach to such a vision is to deny the lower self, putting a taboo on sexuality. In the Orient they teach to befriend the dragon, not to slay him. Eros should become an energy of service to the human community rather than a force of pure seduction for personal gratification. Denial and repression of instincts are the cop-out of the self-righteous. The deer is a spirit of the generative force. Throughout Europe, the deer was commonly portrayed as the sacred ally of the goddess. Deer dancers were part of the New Year celebrations up into the Middle Ages in England and Germany. Deer, for many tribes in the northern hemisphere, were sacred. Their abundance meant life. The Finns went so far as to call the constellation we refer to as the Big Dipper (or Ursa Major) the Great Stag, which shows how important deer were before they were replaced by cows as the major source of red meat.

One of the most memorable moments of my research for this book occurred on a winter day in 1993 that I spent at the home of Winnebago elder Wilbur Black Deer, who has since passed on. Modern science has found that deposits of metal ores in the earth generate electromagnetic fields that can influence consciousness. Nothing happens by chance when intuition rules, and so it should not seem unusual that Wilbur's woodland home is situated next to the quarry in which the largest chunk of nearly pure iron ore was ever mined. Wilbur was a man who had deer power. "Black deer" means "moose" in Winnebago, and Wilbur was a man with special calling for all species of northwoods creatures with antlers. Aside from being a keeper of the old traditions, he was proud that he had helped spur the passage of a Wisconsin law to allow practitioners of the Winnebago religion to take deer for spiritual purposes.

"We have three feasts a year, and venison is a sacrament," Wilbur told me with a smile as I sat with him beside the stove while a light snow was falling outside. Generosity is a mark of the true hunter, and Wilbur and his wife Emily embodied congeniality both to us and to countless members of the tribe during his long life. Outside, his grandchildren and other relatives were preparing a fifty-yard-long lodge out of bent saplings to be covered with tarpaulins for the next feast. Wilbur and Emily were deer people. And in the Winnebago, or Chohunk, culture, deer meat is highly valued. One year, not so very long ago, Wilbur and Emily cleaned and skinned eighty-six deer that people had brought to feasts.

"We have no religion," Wilbur said to me seriously. There was a pause of awkward silence. Outside, a dog barked, announcing another carload of relatives coming for the feast. No religion? I wondered. How could this be? After I had waited long enough to allow the point to be made, Wilbur said, "Our spirituality is our way of life."

The majestic stag with a full rack of antlers is a universal mythic image of potency, but even more powerful, according to ancient wisdom, is the white deer. Albino animals are a genetic "accident," modern science asserts. Perhaps so, but their presence must touch something very deep and precious in the human soul. In Michigan, killing a white deer is illegal.

Sacredness is expressed in many ways. Among the redwood forests of northern California, the white deer is sacred to the Karuk, Hoopa, and Yurok tribes. Just seeing one is thought to bring good luck. Among these tribes, the white deer and the golden eagle are said to have teamed up to create the world. The annual ceremony which recounts the story of creation is the White Deer Dance. Skins of white deer that have been shot or found dead are mounted on poles and carried aloft during the dance, which retells the story of creation, helping a new year get under way.

In the south of the main island of Japan, not far from towering snow-capped Mount Fuji, is the town of Nara, which is known for its sacred deer. As the story goes, one day in ancient times the deity of Kasuga appeared in human form, riding on the back of a white deer. This is an example of the deer as messenger in mythic themes—a common belief around the world.

In Japan, spiritual beings are called *kami*, and according to the old animistic religion, the white deer was a kami. To honor this event, some twelve hundred years ago the Kasuga Taisha shrine was founded in the hills near Nara; at about the same time, the capital was moved to Nara, perhaps because of the omen. The shrine contains many ancient bamboo bows and arrows and a giant drum, as well as numerous effigies of deer, truly a warrior and hunter's shrine.

Today no hunting takes place in Nara. Instead, approximately a thousand sacred deer, descendants of the original messenger deer, roam Nara Park. They are not pure white, but spotted, and they are not shy! Arriving at the parking lot, they wait for your car door to open, then suddenly you have two or three deer poking in your pockets. When you go to the refreshment stand to buy cookies for the deer, they wait in line with you. If you have a good supply, you are soon surrounded by twenty or more spotted deer—which are about the size of goats—including some bucks with antlers, which stand on their hind legs to get the handouts first.

The Bear as Relative

Coming upon a bear in the woods may be a terrifying experience for many people, but in a world where there are no accidents, sighting a bear is a good omen according to my friend, Medicine Grizzly Bear, a Native American shaman and author of the book *Native Healer*. On a quest in the mountains for self-identity and purpose, coming upon a bear or dreaming of a bear signifies a message from the spirit world, for bears symbolize special powers including wisdom, insight, introspection, protection, and healing. Before going into battle, Vikings and other Scandinavian warriors would conduct the *Berserk* ceremony to invoke the spirit of the bear to be with them. Hitler, incidentally, later used the berserk ceremony to inspire his blitzkreigs. Bear are often found near sacred places, which they are said to protect. Artemis, goddess of the hunt, was often depicted as a she-bear. Bears are the guardians of the animal world, nice friends to have on your side.

The cult of the bear is a religious tradition among circumpolar tribes, and wherever it is found the bear is the subject of great respect. Some say this is so because a bear looks more like a human than any other animal. Others say it is because the bear is so large and powerful. Still others note that the bear and the human occupy much the same ecological niche. The reasons for the universality of bear cults are manifold, but one of most notable bear ceremonies is the *Iyomante* of the Ainu, the aboriginal people of Hokkaido, the northern island of Japan.

In 1991, I was the honorary producer of the Spirit of Place symposium in Sendai, Japan. The purpose of this program was to seek bridges of understanding between ancient traditions and modern culture so as to discover better ways to live in harmony with nature. A very special part of that program was the visit of a group of Ainu who brought with them a film which they had allowed to be made of the *Iyomante* ceremony, which honors the bear as their sacred relation.

Anthropologist Carleton Coon classifies the Ainu as part of the Caucasian or white race. They are stocky, and so look more like the Eskimo than the more slender Japanese. Eye color can be blue and there is more facial and body hair. But there has been considerable interracial marriage over the years, and today the Japanese government is reluctant to consider them an indigenous people, even though there are thirty-five thousand registered Ainu living on Hokkaido.

Some native cultures are reluctant to share their customs with others. The reasons for secrecy vary. In some cases, fear of persecution is real. Others feel that people will misuse their ritual forms and knowledge because they do not understand their cultural origins, or because they do not have the proper attitude of respect for the spiritual forces invoked. There have been governmental restrictions on the *Iyomante*, and the Ainu feared that it would die if it was not preserved somehow, so they invited Japanese ethnographic filmmaker Tadayoshi Himeda to film a special ceremony that they would create for his cameras.

Iyomante means "sacrifice," and the purpose of the ceremony is to send away the spirit of the sacred bear, for it is the earth-bound spirit of a mountain god held in a physical body. It is a ten-day rite performed once a year, in winter, presided over by *kamui*, the Ainu god of hunting and the mountains, who is said to live in a large house on the top of a high mountain at the head of the river that runs by the village.

The ceremony involves many elements including brewing a special type of sake using hot stones to speed fermentation; selecting special small trees of certain species from the forest, cutting them off at chest height so they can regenerate, and carving them from the upper trunk to retain delicate shavings so that the finished staffs look like humans; placing these staffs in a line to form an altar, with each staff representing one of the gods of the Ainu pantheon; making a ceremonial bamboo arrow, tipped with a point made from a deer's leg, which is dipped in aconite poison to speed the bear's death; and a yearling bear for sacrifice. The bear has been captured as a cub the spring before and has been raised in the village, treated as one of the humans, including being suckled by some women. During the time of the ceremony, the bear is held in a small cage bound together by grapevines.

After all preparations have been made, the bear is brought out and walked around the village on ropes. When it is time for the killing, the bear is shot with the sacred arrow and then strangled to ensure a quick death. Following this is a feast where the bear himself is a guest, his head being placed in a position of honor at the table while all the bear that can be eaten is consumed, the blood being drunk by men to give them strength. Finally, the skull is placed on a pole and set alongside the others at the altar of shaved staffs, which symbolizes the natural powers of the forest overseeing the ceremony.

According to the Ainu, the world is filled with *kamui* spirits. The Ainu feel that great care must be taken to not offend the spirits. The ritual honoring and killing of the young bear in the *Iyomante* ceremony is performed to honor the bear *kamui*. The Ainu do not feel that they should hunt bears during the rest of the year, but instead, if they pay proper respect to the bear god, then bears are sent to the village by the head *kamui* in the mountains. To keep the attitude of respect, if a bear is killed during the rest of the year, it can only be brought into the home through a special sacred window, and it must be given gifts. The Ainu who came to Spirit of Place asked me to share this ceremony with the world and to teach that the Ainu are a peaceful people who have lived in their ancestral land for thirty thousand years. There are not many bears left, so they do not hunt them much; but they hope that by keeping the *Iyomante* ceremony alive, the peace between humans, the gods, the creatures of nature, and the land will be preserved. The ceremony may seem strange, even barbaric to some, but their legacy of thirty thousand years of survival in this icy cold land suggests that we should not be too quick to judge.

The bear used in the *Iyomante* ceremony is a sacrifice, a term not well understood in modern life. In 1993 the U.S. Supreme Court upheld the right of a Florida church of the Caribbean religion of Santeria to conduct animal sacrifices, the church's attorney pointing out that what the church was doing was certainly no less humane than the slaughtering of cattle for the hamburgers served at the fast food restaurant around the corner from the church.

Santeria is a mixture of Christianity, the Amerindian religion, and Voodoo. Christians speak of Christ's blood sacrifice, and certainly during the history of Christianity millions of people have followed his example in the name of God. In the Voodoo tradition, which is based on rooting nature's powers in human life for service to the community, my West African friends tell me, when an animal is sacrificed, the priest is acting as an obedient agent of a spiritual force beyond this realm. The idea of God calling for blood and death may seem barbaric, but I have seen photographic documentation of a powerful West African priest named Durbach Akuette performing a sacrifice of a chicken, and it seemed that forces beyond the normal were present. The bird was brought to Akuette, who gently picked the chicken out of its cage and held it in the air, offering prayers. Suddenly the bird went limp. It had died of its own accord, seemingly a willing victim.

The Ongoing Legacy of the Hunter

The winter rains have come to the California coast, and snow is falling in the Sierras. Out in the Sacramento Valley, clouds of white snow geese flock, looking for places to feed and rest while hunters hope to entice some into decoy sets made up of hundreds of white rags, cut-out silhouettes, and half-body and full-body white decoys. I doubt if any market hunters are still alive, but each time I come together with other hunters around a glowing wood stove, I cannot help but recall the times I spent as a boy in those musty old garages, front side roasting, back side chilled, as tales of the good old days were swapped. Hunting today is not what it used to be. Some animals have vanished, but others seem to be adjusting to man and will probably outlive man if given half a chance. Death is the fate we all share, but the generative force of nature is equally potent. The hunter, when responding to the primal call of the hunt that is an integral part of our nature, is an agent of balance.

In the final analysis, hunting is a magical practice. It seeks to bring hunter and hunted together. Yet, times change. What is correct for one time, place, and people may not work in another. Hunting is not necessary for immediate survival for most people these days. Nonhunting enjoyment of wildlife is rapidly growing in popularity, while the future of hunting is less certain. If you follow the *Star Trek* television series and movies, you see a future in which food comes from replicators and hunting seems to have vanished from the repertoire of human acts. Activities exist because a culture values them enough to support them. Carl Jung once remarked that bullfighting existed in Spain because it was a symbol of how Spanish men should conquer their instinctual self by being fearless in the face of powerful passions. Watching bullfights was actually the fulfillment of the craving to watch the symbolic inner drama going on in the minds of people of the times, more than a collective lust for blood, Jung believed. But time has passed. Civilization has placed heavier layers of control over the psyche. Instinctual voices grow weaker, and we encourage people to live only in the intellectual, rational mind. Jung suggested that in modern times it might be more appropriate psychologically for people to feed and care for bulls as a symbolic gesture of what was needed to restore sanity to people whose instincts had become so buried.

The snow geese wintering in California's Sacramento Valley were hunted by Alaskan Inuit and other tribes in the north long before the birds arrived here. Subsistence hunting is still very much alive in the far North. In Barrow, Alaska, the last stop on the municipal bus route going north is the shooting grounds for hunting ducks and geese. Among these native people, hunting for food is an essential part of life. Some animal rights advocates go so far as to demand that we should stop all hunting and send canned vegetables to the natives. What we know from research is that this would kill these people, just as surely as giving them blankets with smallpox virus on them. Inuit, Aleut, and Indians of the far North could survive for a time on a vegetarian diet, with vitamin and mineral supplements. It would be very expensive and would likely lead to a rash of diseases in short order, for these people have thousands of years of chemistry that makes them meat eaters. But even more important, if they could not hunt, their lives would lose meaning and purpose, spawning social dis-ease of epidemic proportions. In the aftermath of the 1989 Alaskan oil spill, when food was flown in to subsistence native communities cut off from the sea, there was widespread illness, physical and emotional, just a few weeks after the 11 million barrels of oil gushed into the icy waters of Prince William Sound. The depression and associated malaise among the people was found to be directly related to the loss of both their native food and their lifestyle, which was filled with meaning based on living off the land.

Those of us who are not subsistence hunters face different cultural situations. Our supermarkets provide food. We can get out the camera and binoculars and join the swelling ranks of wildlife watchers. There are lots of ways to get aerobic exercise. I could sell my decoys and make money as a wildlife artist rather than just toss the blocks into the cold waters or muddy fields of winter. I can take my duck call to the marsh and converse with ducks, calling them in to range, but shooting only with a camera. We can get by without hunting, but is this something we really want to do? We could also drop having sexual intercourse in favor of in vitro fertilization. Marriage would not be necessary, so divorce would not occur. Venereal disease, rape, teen pregnancies, and broken hearts could all be abolished by making it illegal to have sex. Perhaps in the near future we could even create ways to have machines nurture children until old enough for what we now call birth. Abortion would be unnecessary, so would cesarean sections. These are all legal and technological possibilities, but what about the human spirit and the great golden web of life? The creation of human self-identity and self-worth can't be taken over by machines. Like all other animals, if we do not pay attention to our instinctive voices we become ill or mad, or both.

Hunting is not what it used to be. One of the new species that both subsistence and nonsubsistence hunters face today is the antihunting activist, a previously undocumented group that today claims a following of 10 million people and nearly $600 million in the combined war chests of their twenty-six most prevalent subspecies. Their cries in recent years have been loud and shrill, and their behavior media-capturing and sometimes destructive. They assert that hunting is a thing of the past, a cruel, brutal, barbaric relic. Hunters, once the heroes who kept food on the table, are now the targets of vicious attacks, their opponents claiming to be the real heroes and heroines. Both sides all too easily call the other "wackos." What do their protests mean for hunting, and society in general?

Before checking out this new category of human, I feel compelled to return to my memories of those old market hunters sitting around the wood stove on cold winter afternoons. Memories of the past give us roots and perspective that can lead to wisdom. If I join my memories with those of the old-timers, I know why I must continue along this path, but before we go any farther, let's look at the opponents of hunting.

Suppressed and wounded instincts are the
dangers threatening civilized man; uninhibited
drives are the dangers threatening primitive
man. . . . Primitive man must tame the animal
in himself and make it his helpful companion;
civilized man must heal the animal in himself
and make it his friend.
ANIELA JAFFE[8]

Three

Animal Rights
and Wrongs

I have known many meat eaters to be far more
non-violent than vegetarians.
MAHATMA GANDHI[1]

Point Reyes is a finger of land that juts into the Pacific Ocean
about an hour's drive north-northwest of the Golden Gate
Bridge. The point itself is part of Point Reyes National Seashore,
the northernmost land on the west side of the San Andreas Fault,
which runs through Tomales Bay just to the east before it slips
into the ocean. The 1906 earthquake that rumbled along the San
Andreas that devastated San Francisco moved Point Reyes sixteen
feet to the west. Point Reyes National Seashore is the closest real
chunk of wilderness to the Bay Area, a dense, mountainous coastal
forest of pine and oaks dripping with moss that suddenly drops
down steep grassy hills onto a sandy beach. The spirit of the wild
is still very much alive there.

When the first white people came to California, they found a
coastal species of elk, the tule, living at Point Reyes and other
areas along the coast and in the open oak forests and marshy areas
of the Central Valley.

In the early 1800s, California may have had as many as 500,000
elk. Explorers reported a single elk herd in the San Joaquin valley
numbering 2,000. As with the buffalo, the twin forces of market

hunting and habitat loss devastated the species. By 1875, there were only two tule elk left in the wild, living in a marsh in Kern County. With careful management, by 1992 the state's tule elk population was restored to 2,500 in twenty-two herds. Like deer, elk are survivors if given half a chance. In 1920, there were an estimated 50,000 elk left in the United States. Today the population nationwide is a million and growing. There are many similarities between deer and elk, but elk are more of a herding animal. Like deer, elk are a good omen, Medicine Grizzly Bear informs me.

In 1978, ten tule elk were introduced into the Point Reyes National Seashore. They were located on a lonely, fog-shrouded, windswept 2,600-acre peninsula on the west side of Tomales Bay and northeast of Point Reyes itself. A seven-foot-high game fence and cattle guard kept them from mingling with nearby herds of dairy cows, for which Point Reyes is also famous.

Except for an occasional mountain lion, there are no predators for the elk at Point Reyes, and by 1993 the herd numbered at least 222, and the count is growing. Concerned about the carrying capacity of the limited area, the rangers at Point Reyes undertook a study of the tule elk herd and its future. Included was a series of public hearings that ran into the fall of that year. In keeping with the spirit of this book, I decided to check out the public report of the final recommendations.

I began college as a wildlife management major. One of the first assignments we were given was to study road-killed animals. While there have been cutbacks on human predation by hunting, the human-driven automobile is rapidly filling that niche of the food chain. As part of our class assignments, we collected recently killed squirrels, skunks, opossums, and raccoons and learned to mount them as study skins. We were also taught to take field notes on roadkills, noting how many of each type of animal were killed per week along certain stretches of road. The number of deaths per mile gives a quick indication of population densities of at least some species, we learned. It also tells you things about the animals. Skunks, for example, are afraid of nothing, for they have no natural predators and so are fearless of automobiles, which makes them one of the most frequent roadkills. Their lack of fear of cars is shared by moose, which also have no serious predators in many areas where they roam these days. In Vermont, where moose are making a rapid comeback, there were 197 auto collisions with them in 1992. Wild turkeys and wild pigs are the two species of common game you almost never see as roadkills. They both are extremely intelligent as well as agile.

On the drive north from my home in southern Marin County, I began logging roadkills. The east-side lanes of Highway 101 north through San Rafael to Lucas Valley Road were pretty clean—I saw the remains of only one fresh deer and one skunk, although a lot more carcasses could probably have been found in the bushes just off the pavement. Turning off 101 and heading west on Lucas Valley Road, changing from a six- to eight-lane divided freeway to a two-lane country road, the distance from nature shrinks. Almost immediately a large turkey vulture swooped in front of my car. I pulled over and noted he'd been having breakfast on a freshly killed black-tailed doe lying about fifty yards away.

In many tribal cultures, the vultures and condors are the agents of the thunder beings who live in the west, creatures who speak lightning bolts of truth through their eyes. Among the Winnebago, Wilbur Black Deer told me, the Thunderbird clan are the peacekeepers. Vultures are messenger birds in shamanic psychology, Medicine Grizzly Bear tells me. Omens, I think, are a personal matter. Sometimes a bird is just a bird, and then there are other times . . .

Winding through the tawny brown Coastal Range, past the gates of George Lucas's Skywalker Ranch, I counted a crow, a towhee, three skunks, a muskrat, and an opossum dead on the pavement or the shoulder. All were just a few days old. Evidence of roadkills usually lasts less than a week, thanks to the scavengers. Just past Nicasio, I passed a truckload of hunters, wearing bright orange, heading up into the hills after blacktail or wild pigs. Like nearly every hunter you'll ever see, they were smiling and happy.

When I arrived at the Bear Valley Visitors Center at Point Reyes, the woman at the desk was a little guarded about disclosing the location where the ranger's report on the tule elk was being presented. I showed her the newspaper story about it and re-assured her that I was a writer and a former professor of natural resources. She breathed a little more easily and pointed toward a white, wooden farmhouse used for meetings.

As I pulled up in the parking lot a good omen appeared! Five feet off the black asphalt lot was an inch-thick red alder tree, the bark stripped to shreds from ground level to three feet up the trunk. Buck rubbing. Point Reyes is crawling with deer! In addi-tion to the native blacktails, in the late 1940s, axis deer from India, which have a spotted coat and long elklike antlers, and ghostly white fallow deer with palmate antlers like those of moose, na-tives of Europe scarcely bigger than a collie, had been introduced by a rancher. The axis and fallow deer escaped into the wild and now are very much at home there, as indicated by the rubbing a scant fifty feet from the administration building. As I mentioned earlier, deer can lose nearly all fear of humans when not hunted.

As I walked to the meeting house, I noticed a vulture that had landed on the fence just across the road, no more than 150 feet away. Sometimes Medicine Grizzly Bear and his omens are a little spooky, but sometimes they are amazingly real.

Inside, there were only sixteen people. Although previous meetings had had some heated debate, according to newspaper stories, everyone seemed very polite and quiet. Shortly, six people filed like a jury into the small room and sat at a table in the front. They were the blue-ribbon elk management committee, headed by Dr. Dale McCullough of the University of California at Berke-ley, who has spent more than thirty years studying the tule elk. A few short years ago, the fate of the elk herd at Point Reyes would have been determined by the park's supervisor at a meeting with California Fish and Game biologists. No longer, since the National Park Service manages Point Reyes, and they generally do not allow hunting on their lands. The panel, which had been flown in from around the United States, had been in conference for a week. The members included university professors, a veterinarian, and three biologists from the Smithsonian Institution.

We soon learned that the present elk range at Point Reyes has an estimated carrying capacity of 350, which should be reached in five years, since there are no predators except the occasional mountain lion that might take a fawn. The tule elk has a very limited gene pool, due to the small size the population had once shrunk to, so it was suggested that a small number of female elk, trapped from other herds, be introduced into the Point Reyes herd periodically to prevent inbreeding, which could lead the herd to extinction.

The question addressed by this panel was how to control the size of the herd so they did not overgraze their range and cause a population collapse along with mass starvation. They reported that there was "no state of the art contraceptive method which could control the tule elk population at Point Reyes." They also noted that it would be possible, eventually, to expand the elk to the entire Point Reyes area, provided certain concerns were addressed. Everyone in the audience seemed excited about that prospect.

One of these concerns was that the elk could have Johnes disease, which is a severe and contagious bacterial infection of the intestine transferred by fecal-oral contact. This is not a disease that can be given to humans, like brucellosis, which can become undulant fever in people, but it could be spread to the nearby cattle ranches that are still within the boundaries of the National Seashore. This is an exotic livestock disease, brought into this country from abroad like brucellosis, which is transported by cattle. Admitting that Johnes disease is very rare among free-ranging animal populations, the panel suggested that the only way to know for sure is to perform a necropsy (an autopsy on a dead animal), and so they recommended shooting fifty animals over the next five years for research purposes.

This recommendation brought out an animal rights activist in the crowd, veterinarian Elliot Katz, president of In Defense of

Animals. The British Royal Society for the Prevention of Cruelty to Animals, founded in London in 1824, is the parent of all modern animal activist groups. Today there may be as many as a thousand animal concern groups in North America with as many as 10 million members. The profile of the average animal rights supporter is a middle-aged white woman with a college education, an annual income around $40,000, who lives on either the coast in an urban area, is a political liberal, supports environmentalist causes, and owns several pets.[2]

Within the animal rights movement is a wide diversity of perspectives and strategies. Some groups that call themselves "animal welfare" organizations tend to be more peaceful, such as the American Humane Association, which runs the familiar shelters that we call "the pound." Others take a more militant position and prefer to be called "animal rights" organizations, such as the Humane Society of the United States, which does not operate shelters. Their positions on hunting are:

> The American Humane Association is opposed to the hunting of any living creature for fun, a trophy or simple sport. . . . American Humane considers sport hunting a violation of the inherent integrity of animals and disruptive of the national balance of the environment through human manipulation, and calls for positive action to prevent such cruelties.

> The Humane Society of the United States is strongly opposed to the hunting of any living creature for fun, trophy, or for sport, because of the trauma, suffering and death to animals which results. . . . The HSUS believes that a civilized society should not condone the killing of any sentient creature as sport.[3]

At the other end of the spectrum is the Animal Liberation Front (ALF), which both Scotland Yard and the FBI have formally designated as a terrorist group. In the United States, the vociferous

People for the Ethical Treatment of Animals (PETA) has frequently publicized ALF's terrorist activities soon after they occur. While PETA itself does not advocate violence and has never approved of ALF's illegal activities, its practice of publicizing such activities has led at least one author to describe PETA as a "PR firm" for terror.[4] And there is talk of an even more violent Hunter Retribution Squad, reputed to have said that the next stage would be to "take a hunter out completely" in retaliation for killing animals.

In Defense of Animals is a vocal group, but it does not encourage violence. Dr. Katz was polite, but proposed studying other animals, such as deer or cattle, and then inferring about Johnes disease in the elk. He said his group does not agree with killing any elk, a point shared by other "anti" groups, and he suggested the need for more research on contraception.

The discussion goes on. The meeting is rather boring, in fact. Then a spokesman for the Rocky Mountain Elk Foundation stands up and asks why don't they consider a special controlled elk hunt for the fifty animals they want to harvest for research purposes. He calls attention to the controlled hunt for tule elk that occurs yearly at the Grizzly Island National Wildlife Refuge, just a few miles away across San Francisco Bay. He adds that some hunters would pay a fair amount of money for the chance to take an elk so close to home, noting that in addition to the open drawing for permits to hunt the Grizzly Island herd, there is a public auction for one tag that may draw as much as $10,000. There is such a raffle for trophy elk hunting in Arizona, which has raised more than $436,000 for habitat restoration and research in five years.

There was an uneasy silence in the room, finally broken by one of the panel members who had been quiet before this. He said simply that the panel agreed a hunt would be a great idea in principle, but that they had decided that if they tried to implement a controlled hunt, all the money brought in by the elk tags and drawings would soon be used up on legal battles with animal rights groups.

With that, the underlying issue of the meeting surfaced: wild-life management is more and more being forced to determine its management policies according to politics rather than science. In the face of legal and political attacks, it has often become a defensive science. When faced with the prospects of violent opposition, a prudent strategy sometimes is to stress the need for more research before coming to any conclusions. Rising costs of research in turn eat away at license revenues, which are a major source of support for all wildlife programs. Then, to try to offset falling license revenues due to the declining numbers of hunters, license fees are raised, which in turn discourages more hunters. As the meeting broke up, a man in front of me said that he'd like to volunteer to be a deputy park ranger. Then he added sarcastically, "That's the only way you'll ever get to hunt these elk."

The turkey vulture was gone as I returned to my truck. The issue has hardly been laid to rest, I could feel it. The endless process of meetings, hearings, proposals, and legal challenges will likely continue. This planning process is not really about elk, it is about a political and philosophical issue that has been galvanized by the elk. Rangers already shoot the exotic axis and fallow deer every year, November through March, to reduce their competition with native blacktails. Animal rights groups don't like this, but don't raise as much commotion about it because these are introduced "exotics," and it is clearly within the mandate of the National Park Service to remove foreign animals and plants from federal park lands to preserve the natural heritage. Whenever I come across this kind of dispute, I like to go to the source, and so I drove north to check in with the elk at Point Reyes.

One dead deer, a squashed opossum, and one flattened skunk later, I had passed through Inverness and was driving slowly through the cattle farms. The country was open. I could see from one side of the point to the other and beyond. All around, the land and air were alive with hunters: red-tailed hawks, two great blue herons after mice in a field; four vultures, six ravens hunting mice in a field as a tribe; a flock of thirteen crows and ravens feeding in a herd of cows; two more flocks of ten ravens and fifteen crows right beside the road; and a number of vultures that floated overhead looking for leftovers.

A seven-foot-high fence marked the boundary of the elk range. The poles were spaced five feet apart, and the fence had four-inch-square heavy wire mesh. North of the cattle guard entrance lie three thousand acres squeezed onto a four-mile windswept spit that is frequently shrouded in coastal fog. The only trees are a few eucalyptus around one old farm building; otherwise the whole peninsula is covered with chest-high brush and grass.

My first greeters from this wildlife sanctuary were a four-point black-tailed buck, a doe, and a fawn. Overhead, two red-tailed hawks drifted in circles and a Cooper's hawk swept by close to the ground. This is mouse heaven, too.

I parked my truck at the first trail. It led out to a sign, describing the elk. Overhead, a sparrow hawk and three vultures watched as I continued along the trail looking for elk. Coming to the top of a hill, I waited for a patch of fog to blow by, then a 150 yards away I saw in the valley below a patch of white. The single most distinguishing characteristic of the elk is its white rear end, which sticks out like a flag. Another term for elk is *wapiti*, the Shawnee Indian word that literally means "white rump." My binoculars confirmed it was an elk rump. Soon I counted fourteen more tule elk in the thicket of blackberries, grass, coyote bush, and coastal sage. The herd bull had six points on either side, and he was stripping velvet. A yearling bull was off to the side. The rest were cows and calves.

I watched the elk for a while. They were peaceful. Thinking back to the hearing, I wondered if hunting could ever become nonconsumptive. One could hunt elk with a tranquilizer gun, have a photo taken with the downed animal for a trophy, and then let it go, like fishing in a catch-and-release trout pond. You could use paint-ball guns to make your mark, but on the other hand elk can get pissed off and kill people with those massive antlers and sharp forefeet. A friend of mine in Michigan who has deer hunting magic—he usually gets a deer with his bow the first week of the season—loves the sport so much that he hunts with rubber-tipped blunts. Claims he gets ten to fifteen deer a year with his blunts, sometimes twice that many. Hunt elk this way in the rut and you'd better have climbable trees nearby.

Off to the left I saw another patch of white emerge from the bushes, so I shifted my binoculars. I thought maybe it was a second bull coming to challenge the herd bull for dominance, or a stray cow finding her way back to the family. The white patch was maybe forty yards from the herd. It was a man's T-shirt. A second white patch joined him. His son. The man lifted his six-year-old son onto his shoulders and began to walk closer to the herd. The bull saw them. The rut is on, you idiot! I thought. That bull could rip you and your kid to shreds, I wanted to scream. The wind was in the wrong direction. If I yelled I would not be heard. There were no trees in sight to climb. The bull was wiser than the man. He gathered the herd together and slipped off into the thicket and the fog.

If you believe that nothing happens by chance, then there is a lesson here. On a previous trip to this range, I met a German tourist with his camera in the parking lot. "Big cats you have here," he said excitedly. He described how he had come within fifty feet of an adult mountain lion, which was probably stalking the elk calves. In places like Yellowstone National Park, bears, buffalo, elk, deer, raccoons, and ground squirrels have become acclimated to people. They are almost a new species. We see pictures of lions and cheetahs sitting on the hoods of trucks of people on photo safaris in Africa.

Life for modern man has become a spectator sport. Television, movies, concerts, magazines, newspapers, books, CD ROM, and video games all provide information about reality. We live our lives at a safe distance from so many things, overwhelmed by information. The animals, however, don't always have the same script. Recently, in one week, reports came across my desk of a logger in Colorado killed and eaten by a bear, a lady in Florida eaten by an alligator, and, in California, a child attacked by a mountain lion, and a man who needed ninety-five stitches to close wounds from a black-tailed buck he had met in his backyard. Other reports described lakes devoid of oxygen in the bottom waters because so many Canada geese defecated in the water, and outbreaks of avian cholera at other lakes filled with ducks and geese. Still other stories told of coyotes breaking into the Los Angeles Zoo and going on a killing spree in the flamingo pond or dragging cats and dogs from the porches of suburban homes in Orange County. In New England, they now have "coy-dogs," coyote-dog crosses, which hunt in packs. These twentieth-century mutts could easily choose humans instead of faster-moving deer. Wildlife management used to mean setting seasons and limits by scientific population studies, planting some shrubs, digging some ponds, checking licenses and bag limits, and looking for poachers. Boy, has that changed! Today, often, more time is spent managing people than working with animals.

I like to consider myself a good tracker. Trailing an animal, you look for signs that will tell you what the animal is doing as well as where it has been. You don't just follow the trail; you have to think a little like the animal to get close. If your minds get into some kind of link, synchronicities can happen. I have been following the trail of animal rights and associated eco-activist groups for some time, talking with people who feel strongly, pro as well as con, and reading material on both points of view. The meeting at Point Reyes, while peaceful, illustrated what is going on across the United States and in many other parts of the world. Hunting is being challenged as cruel, inhumane, sadistic, and barbaric. Typically, all hunters are stereotyped as slobs who get drunk, shoot at anything, never kill cleanly, and leave wounded and dead animals in the woods.

I knew I was getting hotter on the trail when I walked into an office in the financial district in San Francisco and the secretary, who did not know I was working on a book about hunting, said, "Have you seen this?" She handed me the newspaper and pointed to a list of PETA's suggestions for being politically correct toward animals.

The advice included *don't:* eat meat, eggs, dairy products, or honey (bees are smoked to extract honey); wear leather, pearls, wool (shearing hurts sheep), or anything made of silk (worms must be boiled); ride in a horse-drawn carriage; sleep on a down- or feather-filled pillow ("down is plucked repeatedly from the tender skin of living geese"); own a pet, unless it is not subservient to you (don't train it); play baseball, football, or any game whose ball uses animal skin. "What are these people like?" the secretary said, laughing and shaking her head.

I agreed with her—it would be good to know. I knew I was really on the trail a day later when a friend gave me a flyer announcing an upcoming "national wildlife conference" entitled "Whose Wildlife Is This Anyway?" convened by Friends of Animals, Inc., and a flock of animal rights groups. The position of Friends of Animals, Inc. on hunting is:

> The premeditated killing of wildlife is abhorrent to most people, particularly when hunting is condoned under false pretenses, under the guise of "wildlife management, over-population control," or "protection of crops and public safety." . . . *Misinformation* is the tool of special interests, and perhaps nowhere else is it in use as effectively as in the promotion of hunting.

The meeting was to be held in Southern California at the Redondo Beach Holiday Inn on November 6 and 7, 1993. As luck would have it, the week before, a rash of fires erupted all around Los Angeles, including the one that destroyed much of Malibu. Not sure if I was going to be able to get there, I called the conference headquarters to see if the meeting would be taped. "No," I was told coldly by the lady. As I was to soon learn, the tone of the meeting was already being set.

The were ducks, seagulls, cormorants and pelicans feeding in San Francisco Bay around the Oakland airport as the sun rose on the morning of November 6. About an hour and a half later, we could see charred sections of Malibu as the plane was descending, but Redondo Beach was not smoking, at least not visibly. Arriving at the hotel, when I gave my name to the lady at the registration desk, she snipped back, "I thought you weren't coming." I mumbled something and she gave me my name tag and a sample bag of politically correct nonanimal food items. In the hall outside the meeting room were a dozen groups displaying literature and selling items. Bumper stickers seem to be a primary medium of communication among animal rights people. (Hunters usually prefer decals on their car windows that indicate organizational membership rather than political slogans.) Some of the bumper stickers that caught my eye were: HEART ATTACKS—GOD'S REVENGE FOR EATING HIS FRIENDS, SUPPORT THE RIGHT TO ARM BEARS, and LAST CHANCE FOR ANIMALS.

Gathering as much information as I could, I walked inside just as the program was beginning. More than anything else in the room, what caught my eye was the large picture of a bobcat that was on a stand next to the speaker's platform. It was there to call attention to a California Senate bill introduced by Tom Hayden that sought to have the California Fish and Game Commission classify the bobcat as a "game species" rather than a "varmint." The reason for this action is to require environmental impact statements to be filed for bobcat hunting, as EIS are not required for varmint animals. The bobcat is not an endangered species in California. They are common in some areas of the state. The rationale for the bill, as we learned later in the day, was to force the Fish and Game Commission into responding more to nonhunter sentiments. The bobcat photo was not of a peaceful feline but of one with a definitely mean attitude. I could not help but think of Medicine Grizzly Bear's belief that bobcats are a bad omen. They mean that someone is stalking you or going to.

This was not the first time that I had walked into a room filled with people whose views are contrary to mine. In the late 1960s, working with the National Advisory Commission on Civil Disorders, I was a member of a team of social researchers who interviewed block clubs and other citizens' organizations in inner-city Detroit. I spent hours with people whose political views ranged from the far right to the far left. At the University of Michigan, as a student and then faculty member in the 1960s and early 1970s, I watched as waves of social protest came washing through—civil rights, anti-war, ecology, sexual revolution, and drugs. For each movement, student groups were formed, their positions ranging from liberal to conservative and pacifist to violent. In the mid-1970s, I mediated a dispute between the Klamath and Modoc Indians of Klamath Falls, Oregon, and local law enforcement officers. Later in the 1970s, I observed some human-potential and spirituality groups whose ideology verged on cultism. In Seattle, I interviewed prisoners in jail for the criminal justice system. But coming into this room in Redondo Beach, I felt something different in the air.

There were about 150 people. Mostly women. Only two people of color. The audience was exactly as the research on animal rights groups predicts—white, urban, predominantly female, nicely dressed. I saw some older ladies, but none wore tennis shoes. Everyone seemed polite and proper, on the surface, like at a church meeting or maybe a political rally, but there was a tension in the air that suggested a hidden agenda behind the printed one.

The meeting was not taped, so I will report some highlights from my notes. Mark Palmer from the Mountain Lion Protection Fund said there was a "spiritual crisis" in fish and game management going on. He called attention to how hunting and fishing license fees are the primary source of funding for wildlife management and gave reluctant credit to the people who paid these fees for supporting what habitat restoration has been done. But then he noted that there had been a 25 percent decline in hunting licenses sold in California between 1980 and 1989 and voiced his feelings that vested interests were in control of the wildlife of

California. To make his point, he then passed out bags of peanuts to everyone and asked people to send them to Governor Pete Wilson to say that fish and wildlife policies "are nuts."

Palmer called attention to the power of citizen initiatives, including one to secure nonattached public money for wildlife, and ended by defending mountain lions, which have been protected by law as a result of legislation his group got onto the ballot. His presentation was biased but based in fact, and reasoned until he gave his parting shot, which was that sportsmen's groups "will use figures out of context and exaggerate and play games" to protect hunting. He showed he had done some homework and knew how to be polite, but his ultimate agenda was the end of hunting, and he was not above creating a stereotyped enemy to rally against.

Wayne Pacelle, national director of Fund for Animals, was the next speaker. He was introduced as having an environmental studies degree from Yale and being the leader of the Fund for Animals antihunting efforts. Young and well-dressed in suit and tie (almost no one else in the room was so formally attired), Pacelle declared that "the movement" was "fighting thousands of years of prejudice." He compared what was going on with the abolition of slavery, described the history of animal abuse according to Pacelle, including calling Theodore Roosevelt an "inveterate killer of animals" whose behavior was "repugnant," even though he did set aside some lands for refuges—the acreage of which was "very insignificant." Pacelle then listed the species of animals hunted in the United States, as well as how many are killed each year. He made a special point to say that doves "eat weed seeds," which makes them helpful to farmers.

"Animals don't need to be controlled by hunting," Pacelle asserted, failing to note that none of the animals on his list was endangered. "It is not necessary for us to be sport hunters," he said authoritatively, adding that "the majority of Americans oppose sport hunting," (which may or may not be true, depending on whose study you read). I was especially interested in his next statement that "the emotional condition of people is okay."

He went on to report the "Dirty Dozen," or twelve worst hunts in the United States, which the Fund for Animals was trying to stop. This included hunts for certain species such as wolves in Alaska, tundra swans in Virginia, and Texas "canned hunts," youth hunts in Florida, contest shoots of prairie dogs, the use of hounds in hunting bears, hunting stocked pheasant, hunting pen-raised ducks released from towers (100 percent of which are killed by hunters, he said), and hunting "endangered" species such as the Florida fox squirrel. He said the Fund for Animals was going to "take on the gun lobby and the NRA," compared the use of hounds in hunting bears with shooting a caged bear in a zoo, and ended by declaring that "we have reason, logic, and good ethics on our side." The audience loved him.

As Pacelle concluded, people applauded loudly. During the break, he was busy handing out business cards to as many people as possible. Standing in the lobby, I looked out through the large windows, and saw pelicans, seals, seagulls, crows, and blackbirds hunting for food on or along the wharf, which was just a hundred yards away. What is the difference between an animal hunting and a human hunting? I wondered. As I was pondering this, I was approached by two women who wanted to know why I was there. They were trying to look friendly, but they were too pushy to be looking just for polite conversation. I mumbled something and left for the restroom. My purpose was not to be confrontive but to observe and learn.

The next speaker was the burly, charismatic Paul Watson, one of the founders of Greenpeace, who broke off from Greenpeace and went on to form his own organization, the Sea Shepherd Society, which prefers a more confrontive style of protest. Watson described his ramming and sinking of a pirate whaling ship, his scuttling of a Norwegian whaling ship in Antarctica, and other exploits on the high seas where laws are not always clear. "Fear is a cultural construct," Watson declared, noting that he was about to be tried in a Newfoundland court for three counts of criminal mischief for his disruption of cod fishing on the Grand Banks during the spawning season. He said his next boat would be a submarine, so he could spy on people more easily. He gave credit to Cleveland Amory and the Fund for Animals for funding his first ship, which he'd had to sink when he was arrested, but at least he had sold the movie rights, he added. "When you live in a media culture, you have to come up with new ideas all the time," he said, and he called for the formation of "all one movement" to get things done. He received a thunderous standing ovation at the end of his talk.

Heroes do not make themselves; they are made by the culture that endorses and values their actions. In other settings Watson would have been called a dangerous outlaw. People love a "good" outlaw when they lose respect for the law and feel it does not represent them.

The last speaker of the morning was Dr. Michael Klapper, a physician advocate for vegetarianism. He deplored the civilizing of the United States, especially cattle farming, saying that "American agriculture pollutes more than anything else," as there are 1 million cows in the United States. He declared he was "grieved at the destruction of the life force," and felt that because of the violence in the meat production industry, "we are all impoverished for it." He presented some information about the health benefits of vegetarianism, but the success of his attempts to stir emotions ranged widely.

"We are not carnivores," Dr. Klapper said, and advocated growing fiber crops such as jute or hemp to replace our need for trees for paper. "Let the forest come back," he urged, saying that the "single most effective action we can take is to reduce or remove meat from our diet." He then added that it would also be good to hold down family size and drive our cars less.

As Dr. Klapper concluded, Wayne Pacelle came back to the podium and asked people to "take the pledge" not to eat meat, using the coming lunch break to begin a new vegetarian diet. Several times he urged people to take the pledge, and maybe some did. Not I.

I had had enough of the revival meeting and rally. Somehow the thought of sitting in an air-conditioned hotel and eating vegetables (which have no rights, according to Pacelle) did not sit well with my stomach.

I needed some space to clear my head. When I was teaching at the University of Oregon, I used to team with Norman Sundberg, director of the Clinical Psychology program. Our class concerned how to work with human systems, and one of the most effective learning devices we used was a test Norm had created called the Test of Implied Meanings. In this test, people say something with words but underneath their words is another message. Working in counseling, you have to learn to listen for the hidden meanings in people's words, for often they are more important than the actual words spoken. I felt as if the morning's meeting had been like Norm's Test of Implied Meanings.

I slipped out a side door and found to my delight that across the street was Captain Kidd's Restaurant and Fish Market. Since Captain Kidd is a distant relative on my father's side of the family, I thought I should renew the family ties and, in the shelter of my ancestry, try to figure out what was really going on in that meeting. The clam chowder was excellent, and I began to feel contact with the earth again. All around me were happy people downing bowls of chowder and munching on piles of steamed crabs while the football scores were broadcast on the TV. Seagulls on the roof and in the parking lot were there asking for snacks. The restaurant felt so peaceful and friendly.

When I had finished, I walked out to the wharf to the faded blue marlin-shaped sign that announced LANDING. Smiling fishermen carrying burlap bags with bonito, barracuda, halibut, perch, rockfish, sea bass, and mackerel came up onto the dock from a party boat that had just come in. They had spent nineteen dollars for half a day, and most had enough fish for several good meals, not to mention a rejuvenated spirit, which is worth a lot more these days. There is something very satisfying about catching what you eat. The feeling cannot be experienced in any other way, regardless of how much money you spend for food. More than anything, I think it makes you an honest participant in life, and not just a politically correct spectator.

I strolled along the docks, watching seals, grebes, crows, pelicans, cormorants, and willets, all hunting for their lunch. Just north of the Holiday Inn, I noted that there was an electric power generating station that had a mural of gray whales frolicking in the ocean painted on the south side. In the shallow waters under the boats docked in the marina every fifteen to twenty feet there were fish—bright golden garibaldi about the size of a salad plate. Why the even spacing? Territoriality, I learned, as I watched one chase away another who had entered his turf. Nature lays down rules about things. The laws of nature are not always nice, but they work. Thoreau, who supported hunting and was a hunter for much of his life, said that the pattern for human laws should be based on natural laws.

Just what would all these bird, fish, and mammal hunters have to say to the people inside at that meeting? I wondered. Their rights were being discussed, yet who had asked them if they wanted to be represented, let alone if they had any say in this whole thing. The real conference on animal rights was going on outside the hotel, more than inside, I concluded.

Life is such a wonderful treasure, and everything must eat to live. Humans seem so adaptive. But their instinctive voices are weaker than those of other animals, and when they lose touch with their inner voices, they sometimes try to make up rules for right living that may have little to do with honoring their inner nature. What the growing body of research on psychosomatic medicine clearly shows us is that our mental attitudes, more than just our diets, are primary cuases for many of the dis-eases of our times.

Even if you are a vegetarian, to survive you must take the life of other things that have consciousness. Down through the ages, the concept of stewardship—that humans have a responsibility to care for the creatures that feed them—has been the law of the land and it has worked to the benefit of both. They feed us and we care back for them. That, really, is a better deal than other species give their prey.

I checked my watch and headed back to the hotel for more speeches. On the way, I saw two kestrels, small falcons, land on the roof of the hotel. Excuse me if I mention omens and Medicine Grizzly Bear again, but this would certainly be a good omen to him. Two of nature's most deadly hunters had just come to roost on the hotel. Falcons like to kill their prey in flight, hitting a bird with the impact of a pitched baseball and plunging a long talon directly into its heart for a quick kill, if possible. Indians are said to have special knowledge of nature, and I wondered at the meaning of the two falcons showing up. Suddenly the birds swooped down off the roof in a lightning-fast dive for a nearby grove of palm trees. As they disappeared into the trees, a flock of mourning doves erupted out of the grove. Did the mourning doves have any legislation pending in the courts to stop such attacks? Could falcons become vegetarians? Would they take "the pledge"? PETA founder Ingrid Newkirk talks about a day when the lamb and the lion will live together in peace. Obviously the falcons had not been listening through the air vents to the conference inside the Holiday Inn.

The meeting resumed, speaker after speaker sharing statistics, taking potshots and direct shots at hunters, often using inflammatory words to describe hunters—*repugnant, barbaric, abhorrent, cruelty, suffering, murderer, sadistic*—to stir people's juices. The bottom line was well put by the speaker who said, "You have to leave science behind," when you talk about animals, because "their survival is a moral obligation."

For several years George LaPointe of the Louisiana Fish and Game Department has been trying to get animal rights activists and fish and game managers together as part of the Proactive Strategies Project. Using skilled facilitators, LaPointe consistently finds considerable areas of agreement between prohunting and antihunting groups, one of these being the desire to see that no species of game animal goes extinct. Another is the need to stop poaching. A third is to reduce or prevent unnecessary suffering of animals.[5] Professor Jon Hooper of Chico State University has interviewed a large number of animal rights people and also finds many similar areas of agreement. However, when Hooper pointed out these similiarities to the animal rights people, most quickly denied that such agreement existed, he reports.[6]

I stuck it out as long as I could, again declined to take the pledge, and came away feeling that I had been to a rally, at times a revival meeting, but hardly a summit conference on the future of wildlife. There was about as much critical thinking in the presentations as there was meat on the table. People had their agendas and they came in to stomp for support, hunting dollars more than creative answers to wildlife conservation in modern times. It was at times a lively meeting, but the tone was always prickly, like when someone says nice things to you but you know underneath they would love to stab you in the back. Some speakers like Wayne Pacelle and Cleveland Amory were entertaining at times. Both were good at mixing humor with images of violence and inflammatory calls for action, more a clever political fund-raiser's act than that of an honest comedian who gets a check for the gig and walks away not trying to cultivate a cult following. Speakers from fish and game agencies were not there, nor were university professors of wildlife management or even politicians. No one in the room looked like he or she had spent much time outdoors, except Paul Watson. In Alaska they would probably call these folks "ninety-nine fiftiers," which means that 99 percent of them had probably never been more than fifty feet from their cars.

It wasn't a summit meeting of any mountain of issues I could see, but there was an awful lot of hot air, and in the corners were some darker shadows than the lights made. The original literature had advertised special workshops on hunt sabotage. None were on the program, but there were some well-attended special gatherings in the hotel rooms.

Jokes were made about hunting hunters—Cleveland Amory has been doing his "Hunt the Hunters Club" routine for years; its motto is, "If you can't play a sport shoot one"—but people were not speaking out in opposition to taking this motto seriously. When Paul Watson said he had helped Alex Pacheco of PETA get started, people applauded and cheered. Pacheco openly has said that "arson, property destruction, burglary and theft are 'acceptable crimes' when used for the animals' cause."

As I was pulling out of the parking lot, I saw three bumper stickers that seemed to capture a great deal of the spirit of the meeting. One said, SAVE CRUELTY AND CHOLESTEROL. BE A VEGETARIAN. (That is what it said. The grammar is new to me, but I'm a hunter.) The next one said, VISUALIZE INDUSTRIAL COLLAPSE. (Faced with industrial collapse, I envision people hunting like hell to survive.) The third proclaimed, BACK TO THE PLEISTOCENE. (This one I kind of liked—at least there were lots of hunters then.)

The sky was a crimson tapestry and the sun a golden ball slipping into a silvery bank of low ocean clouds. A line of pelicans passed by, looking like miniature pterodactyls, reminding me of the long chain of life that has led us to the present. As I braked at the first stoplight, a new Bronco pulled up next to me, playing rock and roll loud. The real world again! I sighed with relief. Modern protective amulets are commonly hung from our rearview mirrors. I noticed an object hanging in the Bronco. It was a miniature .357 magnum. This was, after all, the Southern California of Rodney King and the L.A. riots. When the officers involved did not receive harsh penalties, the whole area exploded, starting a wave of violence that rolled from coast to coast, revealing the turmoil and the potential for violence that is seemingly everywhere these days. Listen to the news any day of the week—drugs, gang warfare, kids using guns, rapes, shootings, unemployment, homelessness, and so on.

Wayne Pacelle thinks that emotionally, people are reasonably well off these days. I would suggest he consult the January 1994 issue of *Archives of General Psychiatry*, in which University of Michigan professor Ronald Kessler reports from his extremely comprehensive survey of mental health in America today that one-third of the people in the United States are afflicted by mental illness every year, and one-half will suffer from mental illness in their lifetime. The most frequently prescribed drug in the United States today is cimetidine, an antiulcer drug, which has overtaken Valium for the number-one spot.

What about the area of Los Angeles where they filmed *Boyz N the Hood*, which was an arrow shot or two away? What rights do the rats have? Equal to the people, as Peter Singer and other animal rights activists maintain? Tell that to the brothers and sisters and see what they have to say. Farther down the road a black-and-red truck sped past me, the driver a determined-looking bearded man. His bright yellow bumper sticker read, VISUALIZE CIVIL WAR. Two teens standing beside the road caught my attention with the slogans on their T-shirts: LIFE'S HARD, THEN YOU DIE, and, LEAVE ME ALONE!

In the meeting at the Redondo Beach hotel, animal rights was the agenda, but the emotional undercurrent was anger and fear. People talked about animals and cited statistics, but I heard no serious discussions of ways to raise money for habitat restoration and rally support for new and better ways to manage wildlife. As the meeting progressed, pending legislation was identified and people were urged to write to their representatives for or against, without much attempt to help the people think critically about what the legislation was. People were urged to join a group formed to protest others, to give money to the group for reasons that were not clear, and to become vegetarians as a political act. More than a meeting of minds for critical problem-solving on today's serious wildlife issues—habitat restoration and protection, endangered-species preservation, overpopulation of wild animals in urban areas, feral domesticated dogs and cats, and poaching—this meeting, to me, came closer at times to what Freud called a "primitive group."

According to Freud, the "intensification of the affects and the inhibition of the intellect" characterizes primitive groups.[7] People are asked not to make up their own minds based on analysis of the problems but to give up their individual critical thinking and follow charismatic leaders. Political rallies and religious meetings are "primitive groups," according to Freud's terms. So are mobs. Primitive groups are based on strong emotions, passions. So, they don't last long, usually, unless they have an enemy or some other agent to focus on to keep the pot boiling. Hunters seem a convenient target for attention. CASH, the Committee to Abolish Sport Hunting, has declared, "Sport hunting is ecologically and biologically destructive, and is the weakest link in the chain of animal and environmental abuses." Hunters are a perfect scapegoat for channeling emotions from other issues where people feel powerless to one where they may have a chance of seeing results; hunters kill seemingly helpless animals, and, when reason has been set aside, people may more easily identify themselves with the animals. The scapegoating of hunting by animal rights activists is purposeful. "We want to stigmatize hunting," Wayne Pacelle said flatly to *U.S. News and World Report* in a 1990 interview.

Maybe at this stage of the debate about the future of wild animals and hunting it is not possible to bring people together for open debate in a neutral forum. Socialized at the University of Michigan in the 1960s, I am swayed by the Quaker thinking of Kenneth Boulding that teach-ins are sometimes a more useful tool than plain rallies. A teach-in is an open, public debate, allowing all sides to present their positions with equal respect for freedom of speech. Ideally, the meeting is held by an independent organization, that has no stake in the outcome, and no entrance fees are charged. Its power is the chance to destigmatize debate and to inform people and encourage them to think critically for themselves. The teach-in format also offers a way to reduce the chances for violent acting out. If a shooting war breaks out between hunters and animal rights antihunters, I think I know who will win, but it won't solve anything except to determine who is a better shot.

There is much talk about "ethics" in the animal rights movement. Intellectuals, including Peter Singer and Tom Regan, have written very long, passionate essays decrying hunting as unethical for moral reasons. Ethics are among the kinds of regulations that people create for viable community life. They are not cut in stone, except perhaps for those given to Moses, but rather creeds that serve the time, the place, and the people.

The father of wildlife biology, Aldo Leopold, called for a "land ethic" in his poetically powerful philosophy for right living put forth in *A Sand County Almanac*, as we really do not have one for modern society. "A land ethic changes the role of *Homo sapiens* from conqueror of the land-community to plain member and citizen of it," Leopold wrote in 1948.[8]

I am suspicious of people who keep calling for ethics. Real ethics, ones that people cherish, are homegrown. Aldo Leopold, incidentally, recognized this, for he said that just to push out more information on nature and conservation by itself, would never create an honest ethic to the land. Good, useful ethics arise out of long, careful consideration of how human behavior should be in certain situations. They require introspection as well as observation of the world around you and develop from personal experience. Sometimes they are made into laws, other times they remain the unwritten law of the land. Ethics that last and serve a higher purpose obey natural laws and support human health. They are never forced down people's throats or adopted quickly because so-and-so is an entertaining public speaker. Being politically correct does not necessarily make one an ethical person, and often it is a cop-out.

When society is falling apart or injustice is intolerable, people call for ethics to restore order and create peace. The great waves of social protest of the 1960s—civil rights, antiwar, sexual equality, and ecology—were massive social reactions to conditions that people felt were not ethical. The conditions protested against were repressive, people felt. One of the great problems with ethical movements is that they can slip into primitive groups, providing open forums for people to vent unconscious sentiments that may not have much to do with the actual issues. Self-righteousness, frequently a quality of some of the most outspoken ethicists, is all too often a projection of inner furies arising from the fear of self rather than a reflection of the spiritual advancement of the speaker. Truly wise spiritual people are rarely self-righteous. One of the consistent teachings of the sages is to not pay much attention to what people say, but rather to what they do, to know who they truly are. When people decry violence against animals and then support violence against humans, they are not being ethical according to any standard I can comprehend.

There is so much fear now in the consciousness of the people of the United States. The old American dream was held together by a guilt-driven work ethic. When guilt disappears, people will for a time act out to release their pent-up energies—as with the open sexuality of the 1970s. Catharsis may be needed before a new, clear appraisal of the ethical necessities of life can be made. I cannot see how the cause of animals is served by polarizing people, especially when hunters are not the only ones carrying guns. While many hunters own more than one gun, the fact is that there are at least ten times as many guns as hunters in the United States. Some people buy guns for sport, but when people buy guns in record numbers because they feel they need to protect themselves, then there is a lot of fear in the heart of America.

As a social scientist, the Achilles' heel of the animal rights movement that I see is that it is an anticulture culture, at least at this time. Not a counterculture, which seeks to create an alternative society, like a hippie commune, but a culture that bases its existence on needing to have an enemy. People in an anticulture cultural group feel that they have the only legitimate values, that all other views are not legitimate. Violence arises in these situations because they attract people who have other agendas about power and powerlessness, and the angry nature of the group gives them a soapbox from which to vent.

There is money to be made in fueling conflicts between people. We live in a time when all the media scream for our attention by describing and declaring crises as the hottest news of the day.

Lawyers understand this, and so do some social activists. Some make a business of prolonging conflicts, even creating them, to keep paying the bills. They may even expand this mental paradigm into creating whole organizations to fight the scapegoats they have created. Consider the following thoughts by Sam Keen about the psychology of war, and see if they ring true for the animal rights movement and some of its environmentalist supporters.

> Consensual paranoia—the pathology of the normal person who is a a member of a war-justifying society—forms the template from which all the images of the enemy are created. . . . Paranoia involves a complex of mental, emotional, and social mechanisms by which a person or a people claim righteousness and purity, and attribute hostility and evil to the enemy. . . . Paranoia reduces anxiety and guilt by transferring to the other all the characteristics one does not want to recognize in oneself. . . . We only see and acknowledge those negative aspects of the enemy that support the stereotype we have already created.[9]

In addition to the financial benefits of perpetuating an enemy as a reason for belonging, an anticulture offers comfort to others, especially if they feel they have been harmed or wounded by someone else. The wounded animal within now has friends with which to retaliate against someone or something that has caused the hurt. Many have asked if the animal rights movement is not heavily supported by people who have gone through painful divorces or have had traumatic childhoods or have otherwise been hurt by the norms of society. Statistics are unavailable, and it would not be appropriate to generalize, but I can say that many activists of all stripes I saw, as clients in therapy and in classes I have taught, were acting out old conflicts with Mom and Dad as much as trying to help save the earth. We live in times when family values are challenged on many fronts and divorce rates continue to skyrocket. Emotional wounding is tragically common, and emotional education and support are often lacking. My point here is that some people join social causes to act out their unconscious issues, and sometimes these people become visible because they have so much outrageous energy to put into their action.

There is a definite need for a broad-based coalition of groups to join forces to restore the wilderness of America. It is so much easier to criticize than to create. The animal rights movement has a head of steam going. When the movement mushroomed into being, hunters, by and large, were caught off guard. Until the last few years, almost no one had challenged hunting, at least on any organized level. Now, suddenly, many people do. In the last decade the memberships of animal rights groups have increased tenfold. There are still many more hunters than animal rights supporters, but the new advocate groups are a force that cannot be ignored, and whether they will become a positive force for ecological balance, splinter into many different groups and fight more among themselves, or become a revolutionary social movement built on hatred that will add to the violence and climate of fear that seems to permeate all life today, is not at all clear.

There are already some very powerful groups that deal with animal welfare and conservation, including the National Wildlife Federation, the Izaak Walton League, the Sierra Club, the World Wildlife Fund, the Wilderness Society, Defenders of Wildlife, and the National Audubon Society, all of which share a concern for wildlife and none of which are specifically hunting organizations or opposed to ethical hunting. Similarly, hunters' organizations including Ducks Unlimited, the Rocky Mountain Elk Foundation, Ruffed Grouse Society, the Wild Turkey Association, Quail Unlimited, and state groups such as the California Waterfowl Association and the Michigan United Conservation Clubs, are raising millions of dollars for habitat restoration, research and environmental education. The rapidly growing populations of game such as deer, elk, antelope, geese, and turkeys are directly due to their actions.

People often join social movements because of their personal agendas rather than more noble social causes. There are some very real and important animal issues that need to be tackled, but some of the animal rights people are tapping in to issues and energies that go far beyond animal welfare, otherwise they would not be getting such strong support. There is a plague of fear sweeping across America. When people are afraid, they either retreat or attack. With such a diffuse and uncertain enemy to deal with, generalized violence, people look for scapegoats. The hunter has become the target because he or she is easily stereotyped. The reason why the animal rights groups get support is that they are preaching action, not the conventional quiet lobbying or buying up habitat, but here and now drama which gets results. It is like they have learned their tactics from watching bad soap operas and B movies.

It is not necessarily an intense love of animals that prompts individuals to participate in the animal rights movement. Peter Singer, whose book *Animal Liberation* is seen as the jumping-off place for the modern animal rights campaign, openly states that he is not inordinately fond of animals. Wayne Pacelle says he feels the same way, reports Ted Kerasote in his thoughtful study of hunting, *Bloodties*. All these people are effective communicators and political activists looking for support to pay their bills. The animal rights movement is getting so much support because it is also tapping in to our concerns about safety, personal freedom, ecological balance, racial equality, power and powerlessness, violence, terrorism, and the tendency of the media to sensationalize news to compete for ratings. The animal within is in danger, and that is as important to the widespread support for the animal rights movement as is the cause of any furred or feathered creature.

Questions for the Animal Rights Movement

The deer have come down onto our patio again and eaten all the purple petunias, as well as the marigolds, the daisies, and the primroses. Plants with pulled-off stem tops, the characteristic sign of browsing deer, are within two feet of the door. Unless I secure the garbage can lid, raccoons, cats, skunks, and opossums will almost certainly be in the garbage, unless the dogs get there first. Soon we will probably have the same problems with coyotes attacking pets as is commonplace in and around Los Angeles. As more homes are built in wild sections of the country, pets will also be taken by mountain lions. Protected by law, their numbers are increasing, as are complaints of their predation on domestic animals of all kinds and their confrontations with people.

In response to the invasion of animals into my backyard, I could declare war on them, even poach the deer, or be delighted and put out a salt block so they will become regular visitors. My reaction depends on how I feel, as much or more than how much I know. And how I feel about deer, or other issues, is a matter of tracing long lines of emotional energies down into the waters of the unconscious mind. According to the ancient wisdom of many traditions, events occur because of many different forces. There is a spider web of causality to all things. A deer may be just a deer, or it may be a symbol of something else, which has old fires still burning that may not be conscious to the observer of the deer.

America is founded on democracy. Debate is an essential part of democratic society, which is why the First Amendment protects freedom of speech. The animal rights movement has a right to state its case, for whatever reasons it chooses, but it also has a responsibility for the forces that it puts in motion. I hear very little about taking responsibility for their actions in the animal rights literature, when it comes to some of their tactics, and that is a kind of blindness that is very troubling. People can incite violent acts by others. While many animal rights leaders may not be openly calling for murder and arson, their rhetoric and images are often so violent that unstable people can use them to justify violent acts. I personally feel that animal rights leaders share some responsibility for hunters murdered in the Midwest, even if they were not present to pull the trigger.

There is a time and a place for creating polarity in human systems. It makes sparks jump and tensions come to the surface, which sometimes is necessary to resolve a dispute. I learned that from Saul Alinsky, the arch druid of organizers. It is usually easier to organize people against something than in favor of something, especially when fear, anger, and hatred are invoked. It also can result in violence and bloodshed, especially if people feel otherwise threatened by life. Anyone who purposefully seeks to polarize people and incite violence carries enormous personal ethical responsibilities.

The animal rights groups have a following. Money is coming in. Actions are being taken. Media attention is gaining. "The battle" is being fought on many fronts and some victories are being won. But what is the battle all about? Wildlife observers and supporters, by and large, have many things in common with hunters, perhaps many more than animal rights group leaders would like them to recognize. I think the two groups, animal rights and hunters, should be spending more time looking for common goals rather than creating a time- and money-consuming standoff that only benefits attorneys and a few glamour-seeking leaders.

To come to the table and seek ways to help wildlife, I think the animal rights community has to be willing to look honestly at itself and answer some hard questions. Later, I will suggest some things for hunters too, but right now it's the antis I want to address, because they're the ones engaging in bear baiting.

1. Animal rights groups talk about injustice to animals. What is the difference between justice and revenge?

Justice is cold and blind, weighing evidence, seeking balance and peace. Revenge is hot, evil tempered, looking for conflict and blood to avenge past harms, and it tends to result not in peace but in continual conflict and revenge wars. If people can come together in search of higher truths, resolution and respect are possible.

2. Animal rights leaders make frequent use of extremely negative terms to describe hunters and hunting, painting a picture of *all* hunters as dumb, cruel drunks who shoot anything in sight. Do you really believe that all hunters are this way?

3. A lot of the animal rights literature insists that the general public does not support hunting. The surveys that I have seen are

not nearly so black and white. Stephen Kellert has found that a majority of people oppose trophy hunting, and as many as 60 percent of the people he surveyed oppose "sport" hunting, but the majority of the public supports hunting for meat. A survey is only as valid as the questions are able to get clean, correct information. Because of the violent portrayals of sport hunting as evil, the word *sport* may be too loaded now to give accurate results on any public opinions of hunting. Kellert's research was done over a decade ago, when the issue was still young. Some more recent surveys find that a large majority of the general public thinks hunting is okay. A 1993 West Virginia poll conducted by Ryan-McGinn-Samples Research for the *Charleston Daily Mail*, the Associated Press, and WSAZ-TV found that 79 percent of the public approved of hunting. A Minnesota poll conducted by the *Minneapolis Star Tribune* found that 72 percent percent of state residents agreed with the statement that hunting is "a natural activity for people."[10] Other studies show that an equally large number think the tactics of the animal rights movement are not acceptable. So, it is inaccurate to say that the vast majority of Americans do not support hunting.

4. Ethics are what the animal rights leaders talk about, but what are their ethics, and why don't they actively oppose violence among humans? Ethics are only as effective as the people who voice them. Recall the Test of Implied Meanings. Good ethics grow out of personal inner searching for rightness, not from swallowing the views of others. I always have thought that the great philosopher Wittgenstein was right on target when he said that "philosophy is so often the problem which it is supposed to solve." Ethics can be argued forever, and such arguments are heavily biased toward intellectuals, who usually don't hunt. Issues and facts would make better grounds for debate and discussion.

5. Some studies say that 90 percent of the money taken in by animal rights organizations is turned around into more mass fundraising, which is largely based on sensationalism. Others report that animal rights groups are using hostile corporate-takeover tactics of more conservative animal welfare organizations to build

bigger war chests.[11] Is this true? The underlying issue here is whom can one trust, not what is the nature of the animal welfare movement.

6. Many animal rights groups use negativity and sensationalism to bring in contributions. Instead, why not develop self-supporting businesses which use profits from products that adhere to your values? Across the United States, the most genial group events that I have attended are the fund-raisers held by hunting groups. There are no enemies. No paranoia mills are churning. No hate factories. People donate money and sometimes take home items that support the values of the cause, and lots of money is made that goes directly to habitat and animal support. These events are sponsored by volunteers, often at considerable personal expense. No one speaks about being holier than anyone else, family values are promoted, and healthy community life is strengthened. Love of animals and nature, not hatred of people, seems to be the organizing theme of the hunter organizations events I have attended. What would happen if animal rights groups did not target hunters as the enemy?

7. What positive actions are animal rights groups taking to help animals? Let us hear about your habitat restoration work, the sanctuaries you manage, how many researchers you support, what breakthroughs in animal contraception you have come up with, what new animal-free products you have supported the development of, what you are doing to curb poaching, and what new kinds of research models that you have created don't require animals. Nonconsumptive wildlife groups such as the Nature Conservancy and the National Audubon Society run very valuable sanctuary programs that preserve and restore habitat, as well as species.

8. What is the difference between humans killing animals and animals killing other animals? All animals must ultimately die, and natural deaths are not normally quick and humane. Starvation, infection, conflict, and many diseases may include long suffering before death. Some predators begin eating their prey before the

downed animal is dead. Hunters kill animals quickly, sometimes instantly, and certainly as humanely as any other predator. Why is it that killing an animal is so immoral for a person, and not so for a chimpanzee, a cougar, a wolf, or a shark?

9. How do you know that the animals really support you? Among shamanic cultures around the world, in which people seek to establish interspecies communication, there is almost unanimous agreement that animals realize the way that nature works and are willing to let humans kill some of their kind for food as long as they agree to care for the rest of the species. In *Make Prayers to Raven*, Richard Nelson's extraordinary study of the wildlife attitudes of the Koyukon of Alaska, a tribe that has hunted and fished as the primary mode of survival for thousands of years and never created the extinction of a single species, Nelson reports that often the greatest abundance of animals is along the traplines and hunting routes of the shamans.[12]

Setting the Record Straight

Every culture needs its heroes and heroines, who serve as role models to establish and preserve social values and proper behavior. Animal rights people, for example, often claim that their philosophy is rooted in the teachings of Gandhi, Schweitzer, Thoreau, and the ahimsa doctrine of the Jains and the Buddhists. Basing an animal rights code of ethics on such heroes, however, is like building a boat full of leaks. Refer back to the epigraph for this chapter, in which I quote Gandhi's view of the lack of correlation between nonviolence and vegetarianism. Many of the countries where violence is most pronounced are those in which the people are largely vegetarians. Gandhi once killed a cow and reluctantly agreed that poisonous snakes, rabid dogs, or man-eating tigers might have to be killed to save human life. He also felt that "sabotage is a form of violence." I can't believe that Gandhi would join the animal rights movement today.

Albert Schweitzer ate meat, shot poisonous snakes and birds of prey, and commissioned others to kill dangerous wild animals and domestic livestock. He also said that eating meat, at least for some people, was essential to good health and stamina. While he himself was not a hunter, he felt that some hunters were good and moral. Like ethical hunters, Schweitzer understood that a quick kill is sometimes one of the most humane acts of life.

Thoreau fished and hunted and said that hunters had special insights into nature through hunting that nonhunters could not have. In *The Maine Woods* he wrote about deploring the killing of a moose for the pure sport of it, without taking the meat, but admitted that he sorely missed hunting and wished that he could go off and spend a year alone in the woods, hunting and fishing for food. In his later years, Thoreau did become a vegetarian, but in *Walden* he encouraged fathers to teach their sons to hunt and to shoot guns.

Among followers of the ahimsa doctrine (originally a Jain doctrine that means "do not harm" and forbids the faithful from killing any animals and even from farming, which disturbs the soil), there are prohibitions only against ahimsa followers themselves killing animals, but eating meat is not forbidden. Just down the street from me, for example, the Buddhist temple of Marin holds a chicken teriyaki dinner every year as a fund-raiser. Buddha himself considered meat-eating to be "irreproachable nourishment," according to Albert Schweitzer. Incidentally, the Jains created ahimsa not because of any special concern for animal cruelty but because of a desire to lift themselves from this reality and attain etheric consciousness.

Hunters may be under attack, even under the gun, by some animal rights people, but hunters are among the real heroes of our times if you look for the best of what hunting is. Hunters are not trying to sensationalize events to get attention, and yet all across America they are holding meetings and banquets to raise millions for habitat restoration, public education, research, and conducting classes on firearms safety that have made hunting a much safer sport than swimming, football, and baseball, according to the National Safety Council. The hunting literature makes very little attempt to sling dirt at animal rights people, which is in sharp contrast to the hateful diatribes that fill the pages of the animal rights literature.

Hunters go out into the field carrying lethal weapons, but when confronted by angry hunt-harassment parties, hunters show enormous restraint. I know of no situation where a hunter has shot any animal rights hecklers, even though, as the old-timers used to say, "If you want to kill someone, do it in hunting season, 'cause you can always say it was an accident."

In this age when violence fills the nightly news, and behind locked doors people keep loaded weapons for protection, hunters seem a remarkably happy, peaceful group. Probably more at ease than many other social groups. People who keep their cool in the face of such incredible attacks must be rooted in something important to them—something they love. All across America, hunters donate thousands of pounds of food to the needy, spend countless hours on wildlife conservation, and are the primary source of financial support for all the wildlife programs in the United States through license fees and special taxes on equipment. There would not be much wildlife to watch if hunters had not been so good at supporting wildlife conservation programs. The soul of the hunter deserves to be better understood and appreciated. The ethical hunter is a person of the heart.

"The purpose of life," Manor said, "is to stay
alive. Watch any animal in nature—all it tries to
do is stay alive. It doesn't care about beliefs or
philosophy. Whenever an animal's behavior puts
it out of touch with the realities of its existence, it
becomes extinct."
MICHAEL CRICHTON[13]

Four

The Heart of Hunting

All hunters should be nature lovers.

THEODORE ROOSEVELT

In 1903 the Sinclair Oil Company asked my grandfather, James Swan, if they could sink a test well on the southern end of his 550-acre farm on Grosse Ile. There was a chance that there might be oil on the land, they said. My grandfather said "Go ahead."

Having gone down 2,375 feet, Sinclair Oil gave up. They had found no oil but instead had struck a surging artesian well. The water gushed 22 feet into the air when the pipe head was reduced to 4 inches, and the water had a strong rotten-eggs odor from the dissolved hydrogen sulfide gas in it. Because of the water chemistry, some hydrologists suggested its source might be the Canadian shield in northern Ontario, hundreds of miles north.

For a time, the spring drained into a nearby marsh that flowed into Lake Erie. Before it reached the marsh, however, the spring created its own swampy area. The mineral water had a lot of calcium in it, which precipitated out when it hit the air, so all the plants in the water soon were covered with a white, chalky film. The water itself was crystal clear, but everything else in the new pond was white. Snails loved the calcium-rich water, and this in turn attracted flocks of shorebirds—sandpipers, yellowlegs, plovers, snipes, and killdeers—as well as teal, mallards, and black ducks. In winter, the spring never froze; the water temperature remained constant at fifty-two degrees all year.

My father was two years old when the well came in. He still remembers the drilling. Ever since, it has seemed as though he and the spring that feeds the well have had a special relationship. When he was five or six, his father taught him to hunt by giving him an air rifle and teaching him to shoot the tasty shorebirds that were attracted to the white swampy water. His mother, my grandmother, made many a plover pie in those old days, wrapping strips of bacon around the breasts of the tiny birds and baking them in a deep-dish pie crust with gravy and spices.

A few years later the government decided it wanted to dredge out the marsh to make a canal navigable by small boats. Frenchman's Creek thus was born. My father went off to the University of Michigan to play football, but the well's spirit seemed to stay with him, because he studied chemical engineering and later got a license to run water plants.

Then came the Depression and my grandfather tried to sell the farm, but the deal fell through. The flocks of birds attracted to the white chalky outflow provided many meals for the family when times were tough. At times, they were almost subsistence hunters and fishermen.

In the 1930s, my grandfather decided that people might be interested in this water, because mineral springs, like the famous mineral baths in Mount Clemens, Michigan, north of Detroit, were highly valued for their health benefits. Someone from the U.S. Geological Survey came out to monitor the well and found that the daily flow was 2 million gallons a day. A check of the hydrological records revealed that this was the largest artesian well east of the Mississippi. A store and bottling plant were built, and the well water was channeled into a waterwheel to make electricity for the store. Soon the Wonder Well was born, making Ripley's Believe It or Not. Hundreds of people came every weekend to see the well and take away gallons of its water, which many felt was beneficial to health. For those who couldn't stand the smell, a special stream was run through an activated charcoal filter to take away all odors. The Wonder Well remained a major tourist attraction in the Detroit area until the freeways were built, which facilitated weekend get-aways to northern Michigan, and vitamin pills became more popular than mineral water.

Like my father, I learned to fish and hunt where the Wonder Well spilled into Frenchman's Creek. My father would bundle me up in a life jacket and let me wander along the canal and marsh while he waited on water customers, watching me over his shoulder to make sure that I wouldn't drown if I fell in.

I was five years old when I caught my first fish, which was a pumpkinseed sunfish. I used a long cane pole and pounced on the fish when it hit the bank, yelling and screaming like I'd been bitten by a snake. Immediately my father taught me my first lesson about fishing and hunting ethics and conservation: You always eat what you catch and keep. When we cooked the fish for dinner that night, it tasted like oil. The nearby naval air station apparently had been dumping waste oil into Frenchman's Creek, and I learned my second lesson about conservation—water can be polluted and it ruins enjoyment of nature.

The Wonder Well later sent me to college, where, among other things, I studied water pollution and wildlife management. Unlike my father, I decided to tackle the people part of pollution, while he stuck to running the water and tourist business. A decade later, my son caught his first fish, a rock bass, about a hundred feet from where I landed my first sunfish. He too was wearing an orange life jacket that was nearly as big as he was.

Things change so rapidly these days. Ancestry, legacy and heritage are anchors for personal identity that are all too easily overlooked in these days when people are concerned about being politically correct and fads sweep across the United States like winds. Hunting is one of the few areas of life remaining where the experience of modern man is still in sympathy with our ancient ancestors'.

Human beings are not blank pages on which words are written by cultural norms. Each of us has an identity, as well as instinctual cravings that move us to action. If we do not heed the inner energies of our soul, frustration, anger, rage, and illness may result. Psychologist Abraham Maslow proposed that human needs can be ordered into a hierarchy, beginning with the basic immediate needs for food, water, and shelter. Building on this foundation, needs for safety and security come next. When our basic needs are met and we are not preoccupied with concerns about safety, sexual desires, and love, relationships with other people can be pursued, establishing a foundation for personal growth. As our social system begins to work with us, we strive to assert ourselves. As our lower needs are fulfilled, we can turn our attention to becoming who we really are, which Maslow called "self-actualization." Enjoyment, in the final analysis, occurs when our needs are met; joy is experienced as we become ourselves. If we try to deny our instinctual drives, eventually we will cause harm to ourselves or others, for our foundation of identity is not there.

Hunting is not a simple act. Shooting a deer when you are hungry keeps you alive. Making clothes from the skin of the animal keeps you warm. Weaving skins together can create shelter and warmth. Hunting to satisfy basic needs is understandable, especially among people who have little other food, such as circumpolar native cultures. Even among hard-core animal rights activists there is relatively little opposition to subsistence hunting. Among the Inuit, the Saami, the Samoyed, and other tribes, it is not difficult to see how subsistence hunting could lead to realization of higher needs for self-esteem and even self-actualization. Hunters come together to help one another hunt, and then they share meat to create community. Individual hunters choose certain kinds of animals to hunt, and their self-worth is linked with their success at killing, even with the size and number of what they catch and kill. As all the many skills of hunting are learned along the way, mastery emerges, and arts, crafts, and even rituals are created to celebrate hunting, inspiring others to follow this path.

Few modern hunters, however, hunt solely for food, unless they are rural, poor people. I know backwoods roads in the Sierras, in Michigan's Upper Peninsula, and elsewhere, where there is little game of any kind. Just driving down these roads in hunting season sometimes gets you cold stares from people out chopping firewood. These are their deer and rabbits, often jacklighted at night, their defiant eyes tell you. Without them, there might not be any meat on their tables.

Hunting for self-esteem can turn into trophy hunting, which is a favorite target for the antis, who frequently refer to hunting as a "macho" thing and an "ego trip." Hunting as self-actualization seems like psychopathology to some people. Yet, for many passionate hunters, one of the few areas in life which they can feel fully alive is in the hunt. We need to understand the passionate hunter, for it is in the passions that people achieve their highest potentials of being human. How on one hand a person can be a killer of animals and on the other, be a person who is living higher values and putting them into action is a paradox that needs to be understood and appreciated.

For some, hunting is a pleasant pastime; a chance to get out into the woods, enjoy nature, and maybe shoot something. For those who catch "hunting fever," however, hunting becomes a passion. It takes over garages, guides people to paint boats a dull dead-grass color, moves people to purchase expensive wardrobes that often are worn just a few days of the year, and inspires investment in expensive guns, dogs, decoys, calls, and other equipment. In 1991, the 14.1 million hunters who took to the field in the United States spent $12.3 billion, an average of $474 per person for the year on their blood sport, in which they may participate for just a few short days. Many others spend far more to take hunting trips to remote parts of the world—Siberia, Africa, Argentina, Alaska, the Northwest Territories, India, and so on. For the passionate hunter, hunting is not just a sport, or a pastime, it is a life-long pursuit and an expression of self-actualization.

Modern "sport" hunting is a very recent development in human history, and it is quite different from the old ways, not just in terms of the equipment used but in how society views the entire activity. If we focus on the psychology of hunting, however, we find that many of the original elements of hunting remain in the lives of sportsmen today. Each time you take to the fields and forests, there is a long chain of hunters behind you whose collective experience is not all that different from yours. Following the development of hunting through the stages of the human life cycle sheds light on hunting as a legitimate path to self-actualization.

Hunting and the Life Cycle

One of the great psychologists of modern times, Erik Erikson, has proposed that people pass through eight sequential psychological stages in their lifetime, each with its special challenges, strengths, pitfalls, and themes.[1] In early childhood, as a child leaves the mother and begins to explore life's boundaries, the world is a numinous place of odors, sights, textures, and sounds. If pleasant experiences occur, a sense of trust will develop, which will lead to the development of autonomy as well as to a fondness for places that give pleasure. As role models, adults, especially parents, set the stage for developing abilities to trust nature by their behavior in these early years. Simple walks in the woods with your children are an important foundation for the rest of their lives.

If your early experiences are positive, both with your parents and with nature, for the rest of your life you will have a sense of wonder that is nourished by beauty, which allows the emotions of love and hope to emerge. It will be easy to recognize large trees and aged stones as wise and friendly, for they are associated with people who were both. Some people who come from troubled homes adopt nature as a parent or family to replace what they cannot get from humans. The late Chief Justice William O. Douglas, for example, as a teenager "adopted" Mount Adams in Washington State as his "second father," after his father died.

A sense of law and order also begins to develop in early childhood. I wish I could say that the oily taste of my first fish was the only time I experienced that living on Grosse Ile. It was not. Living downwind and downstream of the massive industrial complex of Detroit, we were in the middle of a giant waste stream. Oil spills were a regular occurrence in those days, as were night discharges of wastes into the river and the air. At one time in the 1950s and 1960s the lower Detroit River area was one of the most polluted areas in North America. I am sure growing up there under those conditions has influenced my lifelong commitment to fight pollution. To me, it seems that polluting the air and water is a violation of natural law.

Until the opposite sex and team sports become central issues of adolescence, exploration of nature can be one of the most important aspects of growth in a person's life. Fear is a learned emotional reaction. Infants are startled, but they do not really learn to fear things until they are at least six months old. Fear of the dark and large animals almost never occurs in children under the age of one and a half and usually not before three. After six years of age, children become increasingly resistant to developing new fears of imaginary creatures. Smelling flowers, chasing frogs, watching deer and ducks, catching turtles, and getting wet and muddy are initiations into kinship with nature, which is a tremendous source of health and identity. I suspect that one of the reasons why many young children take such a strong liking to lizards, snakes, toads, turtles, and dinosaurs is that within them the "reptilian" part of their brain is developing and they are unconsciously attracted to animals that symbolize their inner development.

You learn about safety and dangers—deep water, poison ivy, beestings, sunburn, and more—and develop self-confidence from mastering them. If you learn early on that you can trust nature, you will always know that you can return to natural places to be regenerated, no matter what else is going on in your life. This is so because our basic ego nature is symbolized by the forms of nature, and when we are alone in natural places and free of fear, our normal ego boundaries dissolve and the subconscious mind becomes predominant in consciousness. Wilderness then reminds us of who we are, and the natural energies of sympathy charge us with the life force, giving us a "natural high."

You first learn about hunting when you follow grown-ups. As soon as I was able to be quiet, in the evenings after work my father would take me with him to sit in the duck blind not far from where he had learned about hunting from his father. It was a world of excitement and wonder, Dad teaching me to see ducks at a distance, to crouch down as they approached, hearing the booming explosions of his gun, then fetching the dead ducks from the cattails. I never ceased to marvel at how a shot from a gun could so suddenly transform a flying duck, radiating with all the intensity of the spirit of the wild, into a plump, lifeless bundle of feathers.

Erikson believes that "initiative vs. guilt" is a primary issue that children must be taught to understand in their elementary-school years. Hunting and fishing offer an enormous potential to learn life's lessons about initiative and guilt through firsthand encounters with life and death. This is the time when children can learn the basic lesson of life, which Joseph Campbell simply put as "Flesh eats flesh."

In the bible of the animal rights movement, *Animal Liberation*, philosopher Peter Singer wrote, "Practically and psychologically it is impossible to be consistent in one's concern for nonhuman animals while continuing to dine on them."[2]

Many people pick up on this statement by Singer and quote it. The words may be true for him, but his view is not supported by many of the wisest psychologists and the evidence of human history. Carl Jung once said that "goodness which is beyond instinctiveness, which is anti-instinct, is no longer good." Goodness that is based almost solely on the intellect and on book learning and is not linked to our inner nature ultimately becomes destructive, Jung felt.[3]

In support of Jung's observation is analyst Marie-Louise von Franz, who has written extensively about the dangers of modern man losing touch with his unconscious nature. Facing the basic life-and-death issues of being alive, working through the emotions, and coming to a sense of ethical peace with them are an essential part of the process of maturing. Facing painful issues is at the very heart of personal growth. Of the need to accept our need to kill to survive, von Franz wrote:

> Living means murdering from morning to evening; we eat plants and animals . . . plants suffer . . . so vegetarians cannot have the illusion that they do not share in the wheel of destruction. We are murderers and cannot live without murdering. The whole of nature is based on murder. . . . The realization of the destruction and the wish to live are closely connected.[4]

Singer's position is also contrary to the consciousness of hunters and gatherers. The history of man shows clearly that cultures who revere nature, make it the center of their spirituality, tend to depend on hunting for survival. For them, it seems that the need to take life is at the core of their reverence for animals. They turn their spiritual attention to the need to hunt and kill to survive and as a result give us many important insights into how to cope with being ethical murderers. Coming to feel a deep respect, even love, for the creatures they kill is a core issue for all hunters.

Seeing my father shoot ducks and then later eating them, I learned at an early age that meat comes from living things that humans kill. A hunter always hopes that his shots will kill quickly and cleanly. If not, I was taught, it is your duty to find and kill the wounded animal as quickly as possible. When I was a little older, and I had a pellet gun, it was my job to dispatch with a shot to the head the ducks wounded but not killed by the adults. I came to feel very good about this, as I saw that it ended their suffering. It gave me incentive to be a better shot. Still, this was an emotional lesson that took some time. Emotional growth is a slow process, not unlike the fermentation of good wine. Aside from watching my father and other hunters, what helped me overcome my feelings about killing wild animals was to see how wild animals kill one another.

One day as we were paddling the boat out into the marsh, we startled a group of ducks. All but one took off, and it was squawking loudly, flapping its wings, and sinking. As we got close, we saw that a large snapping turtle had grabbed the duck by the foot and was pulling it down for dinner. As it was spring, which is spearfishing season, not duck season, that night we had a wonderful turtle soup and a one legged-duck flew off. (One-legged water- and shorebirds, which are commonly seen, are usually victims of turtles or even large fish. They often live nearly normal lives.)

We had pet ducks, and during winter when the canal froze over we would pen them up, because otherwise they would get frozen into the ice. One time we noticed that we were one duck short. A few feet away I found a half-eaten duck carcass. Soon another duck disappeared from the pen, only to be found in a similar state. The tracks in the snow told the story that a bird had done this. We covered up the top of the pen, and that night saw a very frustrated owl trying to get another duck dinner.

Another time, we awoke and found dead, bloody ducks all around in the pen. Most had little or no signs of being eaten. This time, the tracks told the story that a weasel had gotten into the pen and gone on a killing frenzy. Often we saw how mink would kill muskrats, ripping their throats open and eating just the choice parts of the body. Foxes caught mice, pike swallowed smaller fish, ospreys dropped out of the sky and caught fish hiding under lily pads, herons speared fish, and great horned owls silently swooped down and caught rabbits, all within a hundred feet of my bedroom. Hardly a night went by when I didn't hear strange cries from the marsh, often those of something becoming something else's dinner.

Some predators do not kill according to the ethics of good hunting by humans. Birds of prey sometimes do not kill cleanly in midair, and the wounded birds fall to earth where they cannot be caught. Raccoons, foxes, and skunks love to dine on young ducklings, or grown ducks if they can catch them napping at night.

Wild dogs often take as much as ten minutes to make a kill, disemboweling their prey and eating its guts while the animal is still alive. Hyenas regularly eat their prey alive. Lions sometimes go on killing binges, including attacks on other lions. When George Schaller studied the lions of the Serengeti Plain of Africa, he found their murder rate (lions killing other lions) was nineteen times the murder rate in the United States, which is among the world's highest.[5]

Living in urban areas far removed from wild places and creatures leaves so many children today without the natural way of learning about death, guilt, and eating living things, which is the foundation for developing a healthy reverence for life. To meet the meat-eating demands of the United States, each year American farmers slaughter around 35 million head of cattle, 6 to 7 million sheep, nearly 85 million pigs, 400 million chickens, and close to 300 million turkeys. This all takes place out of sight of most people, so we are shielded from one of life's most important truths— that we kill to live. In the face of an emotionally overwhelming situation, avoidance of troubling situations is understandable. In the long run, however, denial can hinder personal growth.

I never realized how unusual my childhood was until I began teaching human services at the University of Oregon in the mid-1970s. I was hired to develop a special environmental action program for students in the human services. The program was a component of a larger project to staff the "Great Society," which was funded by the Wallace family, who owned *Reader's Digest*. I developed a class called "Environmental Awareness," which focused on the relationship between people and nature. One of the first films I showed was of a group of Eskimo hunting caribou. The caribou were driven into a lake and then hunted down by tribesmen in kayaks, spearing and clubbing them to death. This was by far the most effective way to hunt the caribou without firearms, and my purpose in showing the film was to trigger a discussion about survival needs that all humans share. Whenever I showed the film, invariably as much as half the class became sick to their stomachs. It was the first time many had seen animals die. Only a handful were vegetarians. It was shocking to me to realize that they were that far removed from awareness of their own roots in nature. One of the main causes of the violence and lack of value for human life among children today must be their loss of first-hand contact with the life-and-death processes of nature. The urbanization of life tends to numb us to our connection with nature, and this cannot do anyone good in the long run.

Entering adolescence, people usually first develop bonds with members of their own sex before getting into serious relationships with the opposite sex. In tribal hunting and gathering societies, this is the time when young boys become men through undergoing initiation rituals. It is a time when the cords to the mother are cut and a new sense of responsibility and autonomy is acknowledged. Initiation into adulthood typically involves falling away from the social system of childhood, being stripped down mentally, emotionally and sometimes physically, being exposed to new mysteries, and undergoing ordeals that establish a new personal identity and bonds to a new social group, acceptance of membership in this group, and then a return to community, but as a new person. In many cultures down through the ages, the initiation into a hunting guild or secret society is among the most important times in life.

Among Pueblo Indian children, this period of transition traditionally involved isolating teenagers in underground rooms called *kivas* for up to nine months. During this time they were silent, learning to listen to the voices of the "second mother," the earth. Upon emerging from this ordeal, they were often sent into wild places to pray for visions and dreams that would give vital information about their personal identity and life's purpose. It was during this time that the animal messengers would take the place of the mother as the principal guide. This is the time when our instincts are growing strong, and the bonds struck with animals are crucial triggers of awareness of our instinctual identity.

One of the purposes of initiation into adulthood is to provide a social framework for guiding the use of powerful instinctual energies, such as those used for war and hunting. There are both positive and negative forms of aggression. Defensive aggression is a survival tactic. Offensive aggression starts wars. Hunting is an aggressive act, but according to Erich Fromm there is no evidence to support that hunting is a malevolent form of aggression. Based on one of the most exhaustive studies of aggressiveness undertaken in modern times, including human and animal research, Fromm concluded: "The behavioral patterns and neurological processes in predatory aggression are not analogous to the other types of animal aggression." They are closer, in fact, to the biochemistry of pleasure and joy.[6]

Blood is a key element in initiation into adulthood. Blood is chemically similar to sea water and is the seat of life, soul, and spirit. Women come of age when they start their menstrual cycle. Blood may be shed when virginity is lost. Where circumcision is practiced, it is frequently performed as part of a rite of passage into adulthood. Blood is also sacred. It carries power. The wine of communion in the Christian Church is a metaphor for the blood of Christ. Many hunting rituals involve painting one's body with blood, or drinking the blood of the animals, as it is believed that this conveys the spirit of the animal to the hunter. Among Norwegian hunters, it was customary to drink the blood of bears as this would make the hunters strong like bears. African Masai similarly drink the blood of lions to give them courage. Blood makes us grow up by causing us to face the preciousness of life, and it intensifies an awareness of our own life force.

I first saw a deer killed when I was about thirteen. We were hunting on Drummond Island, a wild gem in Michigan's Upper Peninsula in the St. Mary's River at the head of Lake Huron. In the 1950s, Drummond Island was overrun with deer. It was common to see 150 to 200 during a week in the woods. My father and I were hunting with another father and son. We were driving along a logging road and suddenly four deer walked into a field ahead of us. My father stopped the car. The other man quietly opened the car door, took out his bow, and strung it. The deer just watched. He got out an arrow and stepped a few feet away from the car to get a clear shot. The deer just watched, and the three of us sat in the car watching the drama unfold. He drew back and released the arrow. It sailed fifty yards and struck a spikehorn buck directly behind the foreleg. All the deer bounded off into the brush.

"Just sit tight," my father whispered. "If you don't chase it, the deer will go just a little ways, lay down, bleed to death. Run after it, and it might go a lot farther." The excitement in the air was tremendous. We would have loved nothing more than to run after the deer. But nature rewards patience.

Ten minutes later we walked to the spot were the deer had been hit. The arrow, which had passed right through, lay on the ground, covered with fur and blood. Arrows kill by cutting, severing blood vessels so the animal bleeds to death, unless a vital nerve center like the spinal cord is cut. A trail of drops of blood led off into the brush.

My father made us wait another five minutes, then told us to find and follow the blood trail. Like two dogs, the other boy and I began spotting cherry-red drops on leaves. We slowly worked our way through the brush and saw the deer lying quietly some fifty yards away. "Don't just run up to it," my father cautioned. "Don't know if it's dead yet and you wouldn't want him come after you." Slowly my father and the other man walked up to the deer, nudged it with a boot, and when they were sure it was dead, they let us come. We whooped and ran around like a tribal victory dance! This was a time to celebrate, take pictures, and get excited. In tribal societies, men dance around the bodies of animals they have killed. Like their modern counterparts, they feel joy and thanksgiving for a successful kill. A number of hunters have told me that when they first come upon a deer they have killed, they quietly say prayers of thanks. Some even admit to shedding tears while standing beside a freshly killed buck.

That night back at the lodge where we were staying, the man who owned the place served us fresh deer kidney and heart pie.

But before we could eat any, he had the other boy and me stand in front of the big stone fireplace. He had a little bowl with some blood from the deer, and he painted some on our foreheads and cheeks, saying that we were now members of the deer-hunting society.

Hunting camps are today's versions of the secret societies of tribal hunting cultures, the social groups that men have formed for thousands of years to support the ethics of the hunt and adulthood as well as to have a good time. At their worst, they are drinking societies where little hunting is done and most of the time is spent drinking and playing cards. At their best, hunting clubs are a touchstone of meaning, magic, mystery, and service. Minimally, they restrict access to certain experiences, ensuring that certain standards of hunting are preserved. The best of these clubs invest considerable time, money, and energy into habitat maintenance and restoration, rearing, and releasing of game species, and community service.

On the opposite bank of Frenchman's Creek is forty-acre Round Island, a rich mixture of marsh, woodland, and fields that until recently was a private hunting club which my father and a friend started many years ago. I used to watch after the clubhouse in exchange for hunting privileges and took out guests, acting as a guide. The most memorable was a sportswriter who had never hunted ducks before. He was not a very good shot, but finally when a mixed flock of ducks and coots came in, he dropped a coot. The coot is a slate-gray pigeon-size bird with partially webbed feet and a beak like a chicken. It is a marsh hen, or moor hen, as it is known in some places, for its behavior is more like that of a chicken than a duck. On golf courses all across America these days coots are taking over water holes and leaving a stink behind. I came back to the blind with two scaup, which have bright blue duck bills, webbed feet, and black-and-white bodies, and the coot and laid them on the floor. The sportswriter, a man whose columns on the outdoors were read by thousands of people, looked down and asked me honestly, "Now which is the coot?"

Round Island was a duck-hunting club, a retreat from the world where men went to shoot clay pigeons and ducks, tell stories by the stone fireplace, and sing songs around the old player piano. The odors of the marsh, the cedar walls of the clubhouse and its charcoal grill, the photos on the walls, and the piles of decoys in the shed kindled a sense of high excitement that added richness of tradition to hunting.

You could only get to Round Island by boat or a hand-driven cable ferry across Frenchman's Creek, except when the creek froze over. Every year, after the hunting season, the members held a celebration for their wives, and everyone wore formal evening attire, tuxedos and evening gowns, except for rubber boots. It was a perfect example of one of the most traditional of ritual themes, role reversal, which acts to balance life. Times like this, when normal customs are reversed, such as the beggar king for a day in the Middle Ages, release pent-up energies and keep people from getting too serious about themselves. A hunting club is more than a roof over your head and a place to play cards and drink. It is part of a legacy of secret hunting societies that trace back to the Paleolithic, an island of heritage that serves to focus attention on the magic of the spirit of hunting.

From the earliest times, it has been the job of the hunter to help feed the community. It is the sharing of catch that makes the hunter a hero to others more than his exploits. In keeping with that spirit, today hunters all across America are donating thousands of pounds of wild game to feed the hungry and the homeless. Back in Michigan, many churches are supported by wild game dinners with meat contributed by parishioners. After learning how to kill quickly, the next most important humane lesson for the hunter is to share the bounty of the harvest.

Some animal rights activists charge that giving wild game meat to charities is a publicity stunt, as if it were a new practice. Sharing the fruits of the hunt is actually among the oldest of hunting traditions, perhaps dating back millions of years, and it is found among primates other than man. Jane Goodall discovered that chimpanzees go on hunting trips as often as thirty times a year. Baboons also are very aggressive hunters. According to the noted anthropologist S. L. Washburn, among primates, hunting tends to lead to a more humane social structure than does vegetarianism:

> Hunting not only necessitated new activities and new kinds of cooperation, but changed the role of the adult male in the group. Among vegetarian primates, adult males do not share food. They take the best places for feeding and may even take food from less dominant animals. However, since sharing the kill is normal behavior for many carnivores, economic responsibility of the adult males and the practice of sharing food in the group probably resulted from being carnivorous. The very same actions that caused man to be feared by other animals led to food sharing, more cooperation, and economic independence.[7]

I do not remember ever taking a date out hunting in high school, but on a number of occasions we did organize group outings where several couples went out at night spearing carp. Since this took place during the spawning season, I suppose the symbolism was a little like the Spanish bullfighter who shows that he is a master of primal emotions. One could also make a Freudian argument about the symbolism of the spear being thrust into spawning carp, too. All I recall was that it was a lot of fun for everyone. Later, in college, attending the University of Michigan School of Natural Resources, many women students hunted. It was not the kind of date in which most other students on campus participated, but we had a lot of fun.

Erik Erikson described the basic psychosocial issue of passing into young adulthood as Intimacy versus Isolation. Peer groups are very important, but peer pressure can be deadly sometimes. Drugs, gangs, violence, AIDS, and teen pregnancies are linked to the desire for love and belonging, which is all too often lacking in modern life.

If kids can know that they always have a group of friends in nature, it may be easier for them to say no to those who would drag them into trouble. Gangs develop because there is no other way for some kids to come together and support one another, which may be especially important if parents are at work or families are broken apart. Violence erupts often out of boredom, as do turf battles, drugs, and competition for mates. Competition and cooperation are themes constantly running through young adulthood.

Sports offer an extremely important outlet for many of the tensions of adolescence and adult life. They create special arenas in which natural aggressive energies can be expressed. Where sports have strong support, violence and gangs are less attractive.

People take drugs because they want to feel different. Each drug has its own effect. Marijuana relaxes. Cocaine intensifies. Heroin mellows. LSD opens new possibilities. None is without its dangers. Shamans tell us that drugs are not necessary. There is not a single psychological state that can be reached on drugs that cannot also be attained without drugs.

Getting high is a natural desire, and hunting and fishing are better than drugs if you want to alter your state of mind. They have an element of danger that requires maturity. Life and death are in the hands of the hunter and the fisherman. Success is a mark of mastery. Groups that come together to support hunting and fishing must learn cooperation in order to be successful. Magic, mystery, challenge, and service all result from mastering blood sports. The true lessons of maturity are never forgotten.

Moving into adulthood and middle age, the issues of life involve career and home. The average hunter these days is in his forties, which speaks to a relationship between maturity and hunting that is quite the opposite of the image of the hunter as an immature slob, which is presented by some antihunters. It is also a time when pressures of family and career may pull people away from hunting, however. Hunting today takes more time and money than ever before, as people have less leisure time and the cost of everything has skyrocketed. One of the sad truths of modern times is that many of the best hunting experiences are those that cost a lot of money. If I can afford it, I can hunt for any game I want to. Many places around the world—Siberia, Argentina, Africa, Mongolia, Alaska, and Quebec—are just a few hours away, and guides are waiting there to take me on safaris that resemble those of Teddy Roosevelt a hundred years ago. This is a blessing of modern technology.

When I am on college campuses these days the psychological issue that seems at the forefront of students' minds is fear—the fear of the future. Fear causes us to have tunnel vision and tunnel lives. It is this feeling of fear about modern life and all the troubles of earthly existence that moves people to try to escape; to depart mentally from contemporary reality via drugs, alcohol, and religions that advocate less involvement with the material world. Religions offer comfort, support, and sanctuary from the turmoil of life. Each religion has its lessons to add to the great pool of human wisdom. One of the most important common truths is not to judge others, but one of the sad truths about some organized religions these days is that they have lost touch with their roots in nature. The sacred groves, springs, mountains, forests, and caves that were integral parts of the religious life of hunters and gatherers have become foreign lands to adherents of modern religions. As a result, our lives have become unbalanced, because all these places are reservoirs of spirit. Hunters know this. This is why special places call out to them and they respond to magical sentiments that we have nearly lost our language to describe.

A men's movement has cropped up in recent years in response to the pressures of modern society to turn men into money-making machines instead of allowing them to be human beings. By the tens of thousands, men are going on retreats, pounding on drums, shouting and dancing, and feeling better. This may provide great amusement to watchers, but deep down few people would pass up the chance to let off steam and have a good time.

I have great compassion for many in the men's movement, but I also think that they are kidding themselves if they believe that

being a man stops at pounding a drum. Traditionally, that is preparation for the hunt or the celebration of the kill, not the culminat-
ing expression of masculinity. I recently worked with a graduate
student who started a master's thesis on hunting as a way of studying what he felt to be misguided male energy. He went on an elk
hunt for a week with a group of men. In the process, he suddenly
realized that this was an act of male bonding and of claiming primal power that was far superior to any workshop that he had ever
attended!

Men's mysteries frequently involve themes of blood, overcoming fears, potency, and service—war, hunting, circumcision,
and death. When I see men's groups offering apprenticeships with
master hunters and engaging in habitat restoration work, then I
will feel that the men's movement is coming of age. I have been to
both weekend workshops and weekend hunting parties. I prefer
hunting parties, unless someone pays me to lead a workshop.
Work parties—creating brush piles for small-game shelters, planting trees and shrubs, and making and erecting wood duck-nesting
boxes—are great ways to increase male bonding, too.

In Michigan, a mark of success in life is owning a vacation
cabin in the northern part of the state. Many people prefer to buy
up land in rural areas and possibly move there upon retirement,
rather than invest in huge homes and expensive cars. Similar patterns occur all across the country: People live in cities to make
money, and then as soon as they can they head for the hills.

As we enter the golden years of life, the central psychological
issues are ones of integration and wisdom versus despair. Having
lived six decades or more, people have learned things. Their lives
should work. For many people, this is the time when they can return to hunting, aided by retirement pensions and life savings. On
a recent hunting trip not long ago, my recently retired guide said
that he had worked his whole life just to reach the point where he
could go hunting and fishing every day.

Modern passionate hunters do not hunt solely for meat. Their
needs for food can be met by the grocery store. The modern

hunter receives recreational benefits from being outdoors, getting fresh air and exercise, but there is something more. I recently was talking with a man who said, "I used to hunt. Then I gave it up. Now I just go hiking, and I've lost something."

People today call modern hunting "sport" hunting. *Sport*, according to my dictionary, is "that which amuses in general; diversion; a pastime." There is a lightness to the word *sport* that I think does a disservice to hunting. Like his predecessors, the modern hunter hunts for meaning, to express himself as a member of the human race. Because of the limitations of seasons, distances to travel, financial costs, and shortage of hunting lands, hunting today has become more of a ritual. Yet, hunting remains instinctual in man. Instincts are the driving forces to satisfy our physiological needs. Man has the power to elevate his instincts, to combine them with his intellect. When this happens, man's passions are created. When Mike Billig speaks about his eider hunting as a spiritual experience, he is right on target. The experience of hunting is so special, so different from most other aspects of life today, and so filled with what Jung called "the numinous," that hunting really must ultimately be seen as a spiritual practice by those who sincerely follow the spirit of the hunt. When you visit someone's home and he offers you a wild game meal, know that you are partaking in the sacrament.

One example of the awe and reverence for life that hunters derive from hunting is seen in wildlife art. To be sure, one does not have to be a hunter to paint, carve, or sculpt, but the vital quality of the art of hunters, modern and traditional, speaks of a soul of nature that is the purest and deepest of all human experiences. This is especially the domain of the seasoned hunter, who has learned the lessons of life from hunting and shares their vital spirit in stories, as well as handicrafts. The person who can tell good hunting stories, ones in which all the elements of the setting can be felt as if you are there, is a gift to mankind. Among tribal societies, gifted storytellers are believed to be able to heal people just with their words and rhymes. Such artisans come to their talent not so much by being skilled at speech as by being filled with spirit and able to share their energies and wisdom as a transmission of power.

My father does not hunt anymore. In his nineties, he suffers from arthritis in both shoulders, the result of playing football in college. But he still notes the arrival of each species of duck in the fall, and he can serve up stories about the good old days that will set your mind on fire.

His uncle went deer hunting until he was ninety-six, and his father hunted until he died in his seventies. When people did regular physical labor every day, getting old didn't necessarily mean you gave up hunting. I well remember the last duck my father shot. I had come home to go deer hunting with him. We didn't get anything up north, and so we came back to Grosse Ile and went out duck hunting. It was a bluebird-calm day, and nothing was flying, so I decided to walk around in the marsh to see if I could scare something up. Fifteen minutes later I flushed a flock of mallards and they flew out toward the blind where my father was sitting. Some landed among the decoys and he didn't shoot. I wondered what had happened. I walked back that way and yelled, "Are you okay?" The ducks took off, squawking loudly. Then, one shot rang out. The ducks must have been nearly eighty yards away, but one drake mallard dropped stone dead into the water. I went out to get it. It was an incredible shot, twice the normal distance you would want to shoot a duck.

Back in the blind, my father had a sheepish grin on his face. "I fell asleep," he said with a chuckle. "When I woke up and saw the ducks taking off I felt stupid and so I thought I'd shoot just to keep from looking like I really messed the whole thing up." Then he showed me the spent shell. A shotgun shell fires either many small pellets or a one large slug. For shooting flying birds on the wing, you use shells with many small pellets to create a pattern that makes it easier to get a hit. The larger the number of the shot size, the smaller the individual shots. You use smaller shots, or higher numbers, for smaller birds, to have a more dense pattern. Bigger shot, and smaller numbers, means that individual pellets have more killing power, but there are fewer pellets per shell; size 9 for quail and doves, 6 for pheasant and grouse, 4 for ducks, 2 or BB for geese, and so on. For deer, you use buckshot, which has six large pellets, or one large lead slug about the size of a marble. My father had mistakenly loaded his shotgun with buckshot that we had been using for deer hunting, and one of the six pellets had gone right through the duck's head.

FUNNY THING about the Wonder Well. These days it's just a trickle of water. A few years ago there was a big earthquake in the area and it changed the underground rock strata quite a bit. After that quake, the well slowed to next to nothing. My father had some heart problems the year of that quake, and he had to slow down too. But, just like the well, he still spends some time nearly every day tending business. He used to have an elaborate waterwheel, driven by the artesian spring, to provide all the electricity for the business. In those days when the wheel and the generator were going, the outflow from the well was carried underground by pipe to Frenchman's Creek. As he got older, he took down the wheel, put in some regular electricity, and replaced the tower with a four-foot-high stone pile to support the pipe. Then he shut off the underground pipe and dug out a pond, so that now the chalky white weeds have come back, and so have the snails that love them. The ducks and shorebirds have come back too, just like in the good old days. You can't hunt there today, laws have changed, but the birds flocking to the well sure help to tell some great stories.

As hunters go through life, if they follow the spirit of the hunt faithfully, they are guided through the issues of the life cycle. Times may change, and people may change too, but nature is always there and so are the instinctual energies that call us to hunt. If hunting were not an instinctual act, it could not offer such lifelong potential for self-actualization. It becomes very understandable, then, why as hunters grow older they often become more involved with service organizations that preserve habitat and restore game species. In his youth, Teddy Roosevelt shot wild game all around the world, and the number of kills he made then would not be possible today. Yet, in later years, as he became aware of dwindling animal populations of wildlife and disappearing habitat, he launched the conservation movement. With George Bird Grinnell, in 1887 he created the Boone and Crocket Club, which promoted hunting in "Wild and unknown areas of the country," which became a powerful lobbying group to support wildlife preservation. The Boone and Crocket Club was a model for citizen environmental action soon to be followed by the creation of the Sierra Club in 1892 and the National Audubon Society in 1905.

In contrast to earlier presidents, who did not oppose the military policy of killing off the buffalo in order to cripple Indians, Roosevelt helped establish the first society to preserve the buffalo; and, working with Gifford Pinchot, he created the Forest Service to manage federal forest lands. In 1908, Roosevelt called a National Conservation Congress, attended by most of the governors and some five hundred other political leaders. One consequence of this meeting was that forty-one states soon formed their own natural resource management agencies. During Roosevelt's administration (1901–1909), federal land reserves increased from 45 million acres to 195 million acres, eighteen national monuments were created (including Niagra Falls and the Grand Canyon), five new national parks were established, and fifty-one wildlife refuges were added. Millions of acres and countless millions of wild animals exist today because of Theodore Roosevelt's foresight and love of nature, which can be clearly traced to his lifelong passion for hunting. According to Wayne Pacelle, this isn't much land conservation at all. I guess we'll have to wait until he becomes president to see what a real conservationist does.

Hunters come to know through a lifelong association how precious life is and what a great gift to man is the hunt. Hunting, in the final analysis, is a great teacher of love.

I do not kill with my gun; he who kills with his
gun has forgotten the face of his father.
I kill with my heart.
STEPHEN KING[8]

Five

The Guides of the Hunt

In a very real sense our intellect, interests,
emotions and basic social life—all are
evolutionary products of the success of the
hunting adaptation.

S. L. WASHBURN *and* C. S. LANCASTER[1]

A hunter rises long before dawn, for often the first minutes of the sun's rising, when darkness and light are balancing each other, is the time when game is most active. On icy November mornings when the hunter's moon was still high in the sky, we would head out into the Lake Erie marsh from our home. In many places the water was so shallow that a motor would not work and we had to row or pole the boat by pushing against an oar thrust into the muddy bottom. With practice, one can row or paddle a boat noiselessly. Laws forbid shooting until there is enough light to identify what you are shooting at and to mark where it comes down. But getting from the dock to the blind required passing through prime marshland. And so we came to play a game with flocks of feeding ducks on those predawn mornings: "Can we slip into your flock before you realize we're not a floating log?" On more than one occasion, we poled right into the middle of a hundred or so feeding black mallards. Then suddenly we were recognized and all the world became a whirlwind of feathers, a shower of water and frantic squawking birds, many that we could not even see.

Geese and ducks can fly at forty to fifty miles per hour, traveling both day and night when the migration urge is strong. They orient themselves according to the sun, the stars, and prominent landmarks (including buildings and towers), and the magnetic field of the earth. Research shows that they move in relation to the weather as well as to their needs for food and cover. Many other forces, too, inspire the flight of waterfowl and guide them to where they land; the more we believe that only being rational and "scientific" will guide us to find health and ecological balance, the more we ignore the myriad of forces that move human nature and nature itself. Luke Skywalker called it "the Force."

Ancient Chinese wisdom speaks of the Tao as the flow of life that permeates all things, blending and mixing them together into everchanging new gestalts. Intuition is the human perceptual faculty by which we seek to comprehend the totality, and then to find where we should be at each given moment, when we do not have to follow a schedule or clock.

In all types of hunting, being at the right place at the right time is a key element of success. In the case of waterfowl, finding the correct habitat is essential. Some mallard ducks and Canada geese may now find themselves at home in the small lakes at golf courses, condominium developments, and shopping malls, but most others prefer a wilder setting, and in fact each species has its preferred blend of water, cover, and food.

Bolinas Lagoon, for example, a tidal estuary just north of Stinson Beach, California, is a popular stopover for migrating ducks. On a blustery December day, as many as five thousand ducks of a dozen different species can be seen clustered in flocks around the lagoon. In the deeper channels are goldeneyes, scaup, scoters, mergansers, and cormorants diving for small fish, aquatic plants, crustaceans, and mollusks. Pintails, widgeons, gadwalls, mallards, and shovelers congregate in the shallows, some standing on mud bars exposed at low tide, while others nearby sift through the murky tidal flow for snails, insects, seeds, eel grass, and crustaceans, tipping up so that only their rear ends are showing. Tiny cinnamon, green-winged and blue-winged teal, not much bigger than robins, are gathered in tight flocks where small freshwater streams slip into the saltwater. The shoveler has a broad, flat bill designed for sifting sediment and likes to suck up silt and strain out the tiny critters there, so they flock in shallow marshy areas with fine-sediment bottoms. While dense stands of cattails, bulrushes, and tule provide good concealment for nesting birds, during the rest of the year ducks usually prefer to be out in the open so they can see what is approaching and take flight at the first hint of danger. If the cover is too dense to take off quickly, they are more vulnerable to many four-legged predators. Wood ducks nest in trees but join the other dabblers in shallow-water dining.

Bolinas Lagoon has lots of good habitat for waterfowl, but the birds are not just randomly scattered around the lagoon according to water depth and vegetation. Ducks congregate at certain places year after year—points where the current flow is altered, where freshwater streams flow into the tidal estuary, certain areas in the open water where the bottom has a special abundance of a certain kind of food, and some places that simply seem to draw them like magnets for reasons that man cannot understand. After a time, you come to feel where these places are. Some make perfect sense, while others have a certain air about them that transcends logic. If you have learned how to think like a duck, you know what I mean. The same is true for all other animal species. If you believe in water witching, or dowsing, dowsers say that animals always tend to be attracted to "tracklines," which are places where there is an abundance of positive life force energy.

Learning the habits of animals is a slow, gradual process, a combination of knowledge, awareness, and seat-of-the-pants intuition. For most people the path to mastery of hunting requires time and luck. The best shortcut to the process is to call on the services of a seasoned guide.

Hunting guides are professional wizards—butlers, medics, diplomats, counselors, butchers, game biologists, and more. They are repositories of rare skills including calling and locating game, building blinds, tracking wounded animals, selecting hunting wardrobes, decoy placement, how to keep warm on cold days, and what to do if you fall into deep, soft mud. The answer to the last question, incidentally, is to flatten out as much as possible to distribute your weight so you won't sink as easily. If worst comes to worst, you probably have the best chance not to go under if you can get on your back and do an upside-down breaststroke back to solid ground or the boat. Living along Lake Erie, learning how to get out of soft, deep, black muck is an essential survival skill I acquired early.

Arriving at the blind, it is still pitch dark, but the first task is to set out the decoys. Everything must be in place before shooting starts half an hour before sunrise. Placing decoys is a very precise art. They should all be within shotgun range, which means the farthest should be no more than fifty yards away. Some people will string out a line of decoys leading to the major cluster in front of the blind, creating a J-shaped set. Others have different formulas for spacing, according to species or even to individual decoys. One guide I know always kisses the first decoy he tosses out. Laugh, but he usually is among the first to have his clients limit out. About thirty yards from the blind should be a pocket of open water among the decoys, which is where you want incoming ducks to land.

Ducks of different species will mingle together, but if the water is more than two feet deep, species should be separated. Canvasbacks, redheads, and bluebills should be set be out in the deeper water, for these birds' legs are farther back on the torso, making it easier for them to dive completely underwater to feed. The dabbling ducks, which tip up to feed and have feet closer to the middle of the body, like mallards, black ducks, gadwalls, teal, widgeon, and pintails, as well as geese, do not generally dive, so their decoys should be placed in shallower water, some on land. I like to set off to one side at least half a dozen geese decoys, the larger the better. Even if there are no geese around, the larger-size decoys help attract ducks. The final touch to the set is a great blue heron decoy, placed ten to fifteen yards away. Herons are the sentries of the bird world, and the presence of the heron decoy makes the whole set seem very peaceful and safe.

If you're hunting dabbling ducks in a swamp, you can get by with a dozen decoys, maybe less. If you're hunting Canada geese over water or in a field, you should have at least two dozen decoys, as these are flocking birds. Diving ducks such as scaup, redheads, and canvasbacks live on open bodies of water and often travel in big flocks. I used to use about a hundred decoys when I hunted on Lake Erie. If you go after snow geese or white-fronted geese in places like the Sacramento Valley in California or along the Gulf of Mexico, seasoned guides will use at least a hundred decoys, although most prefer three hundred to fifteen hundred, as these birds are gregarious.

Let's say you are in place, and it's half an hour before sunrise. Shooting is legal. Your eyes are scanning the morning sky, looking for small moving black objects, and your ears are listening for the faint calls of geese. Excitement is in the air, as well as in your body. Adrenaline mixes with the caffeine from your early-morning coffee. You scan the skies for regular fast-moving wing beats on a chunky body flying along in smooth curves or a straight line. The kingfisher has all the makings of a duck, but it makes a rattling call and has irregular wing beats. Herons have wing beats like those of geese, but they pull in their necks and have long legs trailing behind. Egrets stick out their necks, but they have long feet trailing behind. Gulls move their wings slowly, and their wings are more narrow and pointed. Terns have a frantic wing beat and dive frequently. Cormorants look like geese coming in, but they are jet black and clumsy. A pelican is large, like a Canada goose, but the bill makes it impossible to confuse the two. Sandpipers are tiny. Willets and yellowlegs are shorebirds about the size of a teal, and used to be great eating, but they are no longer legal. Their long legs and bill give them away. The long-billed jacksnipe is legal, but it flies so fast and in such an erratic zigzag pattern that hitting one requires an act of God as much as being a good wing shot.

A hunter could blast away at any of these birds, or crows, star-lings, or sparrows for that matter. For the primitive hunter, any-thing was fair game if it could be eaten or used for ceremonial purposes. In certain parts of the world it is legal to hunt the ani-mals we protect. In Italy, for example, songbirds are popular game. Doves are considered songbirds in some states and are the favorite game birds in others. Robins, I understand, are very tasty. Every child learns about "four and twenty blackbirds, baked in a pie," and blackbirds are not usually raised as domestic fowl. Today, hunting regulations dictate what can be hunted, when, and by what methods. For the moment, let's assume that the hunters in the blind are law-abiding fellows.

Off to the left, you see a dozen birds, flying in a tight flock, their wing beats fast and regular. No long legs trailing behind. Necks sticking out in front. They are about the right size and they are coming toward your decoys. You have run down your list of no's and all the rejects are crossed off. Out comes the call. First the hailing call that says, "Hey, come on over." They see the de-coys, veer, and begin to slow their wing beats. They're coming straight in. Now, switch to the chuckling feeding call, "Chow's great down here, come on in." At about sixty yards away the birds veer to one side. Some ducks will make four or five passes before they land. I once sat through eight passes by some very curious canvasbacks before they decided to settle in. The overanxious sky-blaster will lose his cool and shoot here, but the seasoned hunter will hold off and go back to the call, reassuring the birds that everything is just fine, or just sit tight and not move. Motion, more than anything else, will spook animals. Hunting teaches patience, and when you are waiting at times like this, all your senses are running at 100 percent.

The birds have checked things out and are finally coming back, wings set, looking for places to land. When they are just about to land is the optimum time to shoot, for the birds are moving slowly and offer the biggest targets with their wings outstretched. Wait-ing until the ducks are in the water actually makes it harder to get a quick kill because half the bird is under water. But before he raises his gun, the hunter today must also be a game biologist, able to identify species and sexes in flight. In California, in your daily bag limit of four ducks you can have no more than three mallards, only one of which may be female; no more than one pintail of either sex; and no more than two redheads and canvasbacks, singly or in aggregate. If you are shooting at Canada geese, you are forbidden to shoot a cackling Canada goose (*Branta canadensis minima*), which is a miniature Canada goose about the size of a fat mallard, and the Aleutian Canada goose (*Branta canadensis leucopareia*), which has a white ring around its neck. In addition, there are special shorter seasons for certain species, such as white-fronted geese, as well as special regulations within certain parts of the state.

Having checked out the species and sex of the birds that are about to land, let's assume that the birds are legal and desirable. (I will not shoot coots or mergansers as I can't find a way to get them to taste good enough to eat, but they are legal.) Now you can get ready to bring your gun to your shoulder to shoot. Since you are shooting at ducks or geese, federal regulations require that you are using shotguns with shells that have steel shot, not lead, for it has been shown that feeding waterfowl will ingest spent lead shot off the bottom, and this leads to lead poisoning. Steel shot is lighter than lead, so you should wait for even closer shots, if you've been used to shooting with lead. Get ready . . .

The birds are just hanging in the air over the decoys. What do you do now? Neophytes tend to get all excited and start pumping shots off as fast as they can, especially if there are a number of birds in range. Usually this leads to happy ducks and ammunition manufacturers rather than successful duck hunters. The skilled hunter must select which bird to shoot at. This is a crucial point. Your adrenaline is pumping. Your heartbeat is in overdrive. Your hands may be shaking a little. Time seems to slow down. In his profoundly important book about the experience of spirituality, *The Idea of the Holy*, Rudolph Otto says, "To keep a thing holy in the heart means to mark it off by a feeling of peculiar dread, not to be mistaken for the ordinary dread, that is, to appraise it by the category of the numinous." "Religious dread," which Otto speaks of, has the feeling of something "uncanny," "eerie," "weird," and "beyond rationality."[2] Regardless of what you are hunting, elephant or jacksnipe, boar or dove, goose or bear, at the moment when you take aim and shoot, you enter a state of grace, and the experience takes on spiritual dimensions.

Up until this point, there is no difference between the experience of the hunter and that of the nonhunting bird-watcher, who might be using a camera instead of a gun or bow and arrow. The unique experience of hunting unfolds at the moment when the hunter squeezes the trigger or releases the bowstring for the anticipated outcome is the death of the animal.

The life-and-death decision being made by the hunter to shoot or not is the sort of situation that calls upon many faculties—intuition, attitudes, values, and ethics—and then requires you to blend them into a unity of mind, body, and spirit for action. A quick, clean kill is what every hunter should be hoping for, even praying for, and the chances of this happening will be greatly increased if the hunter is at peace with the forces that guide him.

Challenges to Hunting

For 99 percent of human history, and even more so with modern weaponry, the automatic reaction for the hunter when he sees ducks forty yards away is to start shooting. Action would be a matter of instinct, and hitting a moving target is instinctual shooting, not mechanical sighting. The best shots are ones who can set aside their conscious minds and shoot from their center, with bodies, minds, and weapons becoming a lethal unity. Wait! Perhaps we will get a chance sometime, but right now something has intervened. From behind us there are sounds of people screaming and beating on pots and pans. The ducks veer off. No shot is possible here. From out of the tangled thicket of willows and alders behind us emerges a hunter harassment party.

There is now a standoff. The animal rights activists are screaming insults and taunting the hunters, "Cruel beast!"; "Murderer!"; "Can't you prove your manhood in bed?"; "Sadists!"; and, "Those helpless ducks, you should be ashamed of yourself!"

The temptation runs across the hunters' minds that these humans could be the perfect substitute for the ducks that have just flown away. It's an honest thought, but I know of only one case of a hunter shooting in response to such harassment, and he shot over the protester's heads. The restraint shown is certainly aided by the requirement in nearly all states that every new hunter must complete a course in hunter safety, which includes passing a written examination, before he or she can get a hunting license. Some hunters I know carry cans of mace, used to repel muggers, in case they are confronted by harassers. A few carry cellular phones to report hunter harassment, which is illegal in nearly all states. The Recreational Hunting Safety and Preservation Act of 1993 makes hunt disruption on national forests, public lands, national parks, and federal wildlife refuges a federal crime punishable by a civil penalty of five hundred to five thousand dollars for each violation, and if violence is involved, civil penalties of one thousand to ten thousand dollars for each violation, in addition to any other criminal or civil penalties.[3] Assume that the call has gone in to the 800 hunter harassment number most states now have, or to the local 911, and that the police and/or game wardens are on the way. The ducks are gone, and so is the hunting for a while. Let's see if we can stage a debate to hold their interest while the cops come.

If the harassers are not carrying guns, or have not shot paintballs at the hunters, thrown things at them, or committed acts of vandalism against their boats, cars, or dogs, any of which could push a hunter to act in self-defense or make a citizen's arrest at gunpoint, let's look at some of the typical antihunting accusations presented below and give the hunters a chance to respond. These antihunting positions are based on the literature of antihunting groups and their spokespeople. The following is what an interchange between the two sides might sound like if we can go beyond the baiting and taunting stage.

Antihunting accusation: Only 10% of the population hunts, yet the rest of us pay for the state and federal wildlife agencies that artificially increase wildlife populations to satisfy hunters. Hunting licenses account for only a fraction of the cost of hunting programs. Overpopulation (of wildlife) is an unnatural condition created—by our tax dollars—for the benefit of hunters. (Source: The Animal Rights Handbook [Venice, CA: Living Planet Press, 1990], pp. 83–84.)

Hunting response: The U.S. Fish and Wildlife Service is concerned with management of *all* species of fish and wildlife. Hunters may legally pursue only about 145 of the more than 1,150 species of birds and mammals in the United States. The work of the more than 6,500 employees of the Fish and Wildlife Service includes operating more than 700 field units and managing more than 450 refuges, the world's largest refuge system (started in 1903 by Theodore Roosevelt). Other projects, including research programs on pesticides and their effects on wildlife, census counts of birds and mammals, environmental education, and work to save endangered species, consume a great deal of the resources of the Fish and Wildlife Service.

In 1991, of the 108.7 million Americans who participated in wildlife-related recreation, hunters represented about 17 percent of this total, yet hunters and fishermen are in most cases the only recreationists who actually pay for their privileges of wildlife-related recreation. Protection against animal rights actions, including lawsuits and sabotage of research and law enforcement to curb hunter harassment, incidentally, is a rapidly growing budget item in federal and state agencies, consuming money and resources that could be used to help the animals.

Hunting license fees contribute about 50 percent of the revenue for state fish and wildlife management agencies. License sales have generated close to $3 billion since 1923. Every hunter over the age of sixteen who hunts for migratory waterfowl must buy a federal duck stamp, the moneys from which go to federal waterfowl research and conservation. Since 1934, duck stamp fees have brought in close to $200 million. Also, most states now have a state duck stamp, and some have special stamps for upland game, as well as special license fees for big game, which are in addition to the regular licenses every hunter must have. Such stamps contribute about 18 percent of the funds for wildlife management agencies.

The Federal Aid in Wildlife Restoration Act (Pittman-Robertson Act) of 1937 places an excise tax on sales of firearms, ammunition, and archery equipment. Of the moneys raised since the law's passage, over $1.5 billion, 62 percent is redistributed to states to buy, develop, and maintain wildlife management areas; 26 percent goes to the states for surveys; and 7 percent goes to support hunter education classes. These moneys represent about 23 percent of the overall revenues for agencies.

The bottom line is that nearly 75 percent of the annual income of state fish and game agencies comes from hunters and fishermen, according to the U.S. Fish and Wildlife Service, the Wildlife Conservation Fund of America, and the Wildlife Management Institute. We could use a lot more money to manage fish and game, but so far the support of fish and game agencies has been shouldered mainly by sportsmen. And we should add that sportsmen contribute many millions more for research and habitat restoration through conservation organizations such as Ducks Unlimited, Trout Unlimited, and the Rocky Mountain Elk Foundation.

In one breath animal rights activists talk about endangered species and then in the next charge that hunters and game commissions create overpopulation, especially of deer, but also of pheasant, ducks, geese, rabbits, elk, and other animals. Overhunting of predators, such as grizzly bears and wolves, in some areas did boost deer populations in the past. But today, overpopulation of deer and other species, which is common in many areas of the United States, especially the Northeast, is more related to restrictions placed on hunting. All across America, when state regulatory agencies try to increase the kill of deer, especially taking more antlerless deer, animal rights groups howl their protests, and bigger and bigger bureaucracies are created as a defensive measure.

If all hunting in America were to stop tomorrow, the consequences for many species would be devastating. In short order roadkill crashes would skyrocket; crop predation would rise; outbreaks of disease among animals would jump dramatically, in some cases causing large die-offs and massive suffering; negative encounters between people and animals would increase; the incidence of rabies would rise; Lyme disease would become an even more widespread public health menace; some habitats would be destroyed; and the populations of many species, such as deer, rabbits, and Canada geese, would rise dramatically. In time, a new self-limiting population would be established, but there are no guarantees about what would go on between now and then. In the meantime the suffering of the animals would be far greater than it is now. The most likely predators to increase in many areas would not be wild animals but feral cats and dogs; and feral dogs are also a threat to humans, for they are the least prepared of any species to survive in the wild and they can learn to hunt in packs.

Antihunting accusation: Each year hunters kill an average of 400 people. (Source: Anna Sequoia of Animal Rights International, in *67 Ways to Save Animals* [New York: HarperCollins, 1990].)

In 1988, 177 people were killed and 1,719 injured, many innocent bystanders walking in the woods or on their own property. (Cleveland Amory, president of Fund for Animals, quoted in *U.S. News and World Report*, 5 February 1990.)

Hunting response: Aside from the fact that there is a difference of 233 deaths from hunting between the charges of Anna Sequoia and those of Cleveland Amory, the National Safety Council reports that for 1988 there were 161 hunting fatalities, 49 of which were self-inflicted. Thanks in part to hunter safety education classes, hunting fatalities have declined by more than 50 percent over the last two decades.

In general, three-quarters of the hunters who have accidents have not taken a hunter safety education course. Participating in hunting today is safer than swimming, bicycling, playing baseball, golf, tennis, touch football, basketball, fishing, horseback riding, and driving to the place where you hunt, if you look at the number of injuries per 100,000 people participating in various sports compiled by the National Safety Council. In 1988, ten states reported no hunting fatalities, and Connecticut had no hunting accidents at all. Statistics show that you are more likely to be killed by lightning when outdoors than to be killed in a hunting accident. In a normal season, more hunters die from heart attacks than hunting accidents. According to the California Department of Fish and Game, there is a 0.0015–0.00425 percent chance of being killed or wounded while hunting deer in California. In 1992, despite the presence of nearly half a million deer hunters in the field, no one was killed and only one person was wounded in California.

In response to Cleveland Amory's charge that hunters are harming "many innocent bystanders," the actual data show that "Hunting accidents involving non hunters are extremely rare. On the average, only one nonhunter is injured by a hunter for every 12 million recreation days of hunting. A nonhunter is 20 times more likely to die from stinging insects than wounding by a hunter."[4] Media tend to sensationalize accidental hunting deaths and injuries, but in comparison with many urban areas where violence has reached epidemic proportions, the woods and marshes during hunting season are extremely safe, especially when you consider that everyone hunting is armed with lethal weapons. In 1992 in California there were no nonhunter injuries or deaths associated with hunting.

Antihunting accusation: The premeditated killing of wildlife is abhorrent to most people. (Source: "Hunting: An Act Against Nature," Friends of Animals publication, n.d.)

Hunting response: Stephen Kellert's research in the late 1970's and early 1980s found that about 60 percent of the general population was opposed to sport hunting, but only 20 percent was opposed to hunting for meat. Sixty-four percent of Kellert's respondents felt that hunting for meat and recreation was acceptable.

A poll is only as good as the questions asked and the sample polled. The sample who is also asked questions about animal rights is especially important as pro and con views vary considerably from region to region. Urban areas, especially on the West Coast, seem to have the highest opposition to hunting, while rural areas in the Midwest, South, and Northeast have the greatest hunting supporters. A *Los Angeles Times* poll in December 1993 reported that 54 percent of the respondents opposed hunting for sport. Since most polls have less than a 5 percent error factor, one could argue that support for sport hunting has slightly increased since Kellert's research. This would be the case especially if the *L.A. Times* sample was done in West Coast urban areas, which tend to have the most antihunting advocates. Most national studies show about a 50–50 split in prohunting and antihunting opinions, but the results are significantly influenced by the actual words used in the questions.

The use of the term *sport* increases bias in surveys. It tends to portray hunting as a sort of lethal tag game, and conjures up memories of the 1800s when buffalo were shot for pleasure from trains or passenger pigeons were caught and released as live targets for shotgunners, or images in modern times of a bunch of drunks out in the woods shooting at animals for target practice. An example of how people respond emotionally to the word *sport* is the statement of the American Humane Society that it "believes that sport hunting is a form of exploitation of animals for the entertainment of the hunter, and is contrary to the values of compassion and respect for all life that inform the Humane Society's mission." Or the position of the Humane Society of the United States (which does not operate dog pounds), that "A civilized society should not condone the killing of any sentient creature as sport."

In 1993, about half the population of Arkansas bought hunting or fishing licenses. Michigan sends 750,000 hunters to the woods every year. A poll on hunting conducted in states such as Minnesota, Wisconsin, and Pennsylvania would probably find much more support for hunting, as would be the case in Louisiana, where the license plates carry the state motto SPORTSMEN'S PARADISE. In Boise, Idaho, a popular bumper sticker is GUN CONTROL MEANS USE BOTH HANDS. At a shopping mall in Boise, the largest store is Intermountain Arms and Tackle; it's bigger than the mall supermarket. Opposition to hunting is strongest among people who live in urban areas and those who never had the opportunity to go hunting as children.

A 1990 Gallup Poll of one thousand Americans asked the question: "Certain animal rights groups want a total ban on all types of hunting. Do you support or oppose this goal?" Of the respondents, 77 percent opposed this goal. The poll found that 72 percent did have at least some respect for the positions which animal rights activists take, but 90 percent opposed hunter harassment. And when asked about their overall attitudes toward animal rights activists, only 7 percent agreed with what the animal rights groups are trying to accomplish and approve of how they are going about it.[5]

Antihunting accusation: The interests of animals ought to be considered equally with our own interests and from this equality it follows that we ought to become vegetarians. (Source: Peter Singer, "Killing Humans and Killing Animals," *Inquiry* (22): pp. 145–56.

Hunting response: Singer, who considers eating meat an act of "speciesism," or one species discriminating against another, is making an argument about diet based on ethics, which may or may not have anything to do with health. If you can stay healthy on a vegetarian diet, great. Many people can't. I'll invite a guest "expert witness," the renowned chef Julia Child, to respond to Singer. In a

1991 *Newsweek* interview, Child stated: "I'm seventy-eight years old. I work hard. I'm up at six every morning, and I need my strength. I feel psychologically and spiritually that I need meat. . . . I want to be healthy and well fed up to the end. What will prolong my life is eating well and enjoying it."

Julia Child's attitude is very healthy from a psychological standpoint. When we care for our true self, this is healthy self-love, and such a person will be happier and healthier than one who practices injurious self-denial. Diet based on political views is a statement of self-righteousness and martyrdom more than self-affirmation. This position will frequently result in some kind of anger clouding one's perception of life, inspiring violence from a person who supposedly supports peace. The violence is due to the conflict between bodily needs and a behavior imposed by a judgmental super ego. If people eat meat because it is good, even essential, for their health, then it would seem that people who oppose them would be guilty of dietary discrimination.

The research clearly shows that hunters become especially fond of animals they hunt and nature in general. Anyone who argues about the rights of animals should be able to search through their conscience and feel that they have the support of the animals on this matter. After all, the animals should have the right to choose who represents them. In Richard Nelson's studies of hunters and trappers of the Yukon, reported in *Make Prayers to Raven* and other writings, he says that animals are often most abundant along the traplines of shamans, who supposedly have the abilities to converse with animals.

Antihunting accusation: According to Wayne Pacelle, executive director of the Fund for Animals, writing in Animals' Agenda, in 1987 hunters killed and crippled 10 million ducks (whose numbers are now at dangerously low levels). (Source: Sequoia, *67 Ways*, p. 38.)

Hunting response: As of spring 1993, the duck population of North America is about 60 million, which is down from a mid-

1970s count of close to 100 million. Duck populations are dependent on spring water levels, and between now and then there have been some serious droughts. According to Ducks Unlimited, which mans its own major research effort and works with wildlife management agencies in Canada and the United States, "only about 8 percent of the fall flight of ducks is killed through hunting (including ducks shot and crippled but not found and killed, and illegal harvests)." To fully appreciate this statistic, one should also know that nesting ducks have only about a 15 to 20 percent rate of success due to predation by other animals, which is a much more important factor in the killing of ducks, and feral domestic animals take an increasingly large number of ducklings every year. At least 20 million ducks also die annually from diseases including avian cholera, botulism, DVE (duck plague), and avian influenza. Such diseases occur and are transmitted most easily when ducks are crowded into restricted areas, which is increasingly common these days due to habitat loss and overpopulation, especially on lakes and ponds in urban areas where no hunting occurs. Add to this deaths due to weather (ducks frequently freeze their legs into water at night and perish there, and hailstorms can kill ducklings and eggs), poisoning by pesticides, subsistence hunting by native tribes for eggs and adults, and collisions with wires, trees, buildings, cars, and the like. It is interesting to note that the survival percentage rates of ducks do not change much from year to year, even though the actual numbers of birds may vary. When there are fewer ducks, there are soon fewer duck hunters. Hunting seems to be self-regulating.

A number of studies show that if duck hunting is stopped, duck populations will not rise much, because predation and disease will take more ducks. The same seems true for ruffed grouse. A six-year study of ruffed grouse populations in Michigan conducted in the 1950s compared populations of grouse in two areas with identical habitat, one heavily hunted and one protected, and found that there was not a significant difference in population between these two areas.

If the antihunters considered the facts more carefully, they might join the hunters to curb pesticide use and save wetlands. As long as wildlife biologists keep a good count on populations, hunting will never be a factor in the extinction of any species. Since market hunting has been abolished, neither hunting nor trapping has been responsible for the extinction of any species of bird or mammal in North America. Man has been a natural predator for hundreds of thousands of years and has as much right as any other predator to hunt for food.

Antihunting accusation: The late Dr. Karl Menninger, Sr., ascribed hunters' "joy of killing or inflicting pain" to having "erotic sadistic motivation." (Source: Sequoia, *67 Ways,* p. 39.)

Hunting response: Stephen Kellert studied antihunters' attitudes as well as those of hunters, and he found that many people whom he called "moralistic anti-hunters" consider hunters to be motivated by psychopathological impulses.[6] This quote by the late Karl Menninger is frequently used by moralistic animal rights activists. The earliest use of the quote I can find is in the 1975 animal rights book *The Politics of Extinction* by Lewis Regenstein, which is almost identical to the Sequoia quote, and its source is not given. Using this quote and an undated letter from University of Michigan instructor of neurology and neurosurgery Dr. Joel Saper, Regenstein offers a "Psychology of Hunting" that claims, "Some psychologists and psychiatrists feel that people hunt as a compensation for a lack of 'masculinity.'"[7] Saper is free to have personal opinions about hunting, but the areas for which he is well-known are anatomy and physiology, not the psyche. Menninger, however, was a well-known, well-respected psychiatrist, so I will address his remarks in some detail.

Karl Menninger was an outspoken advocate of humane mental health treatment programs and an equally outspoken critic of the penal system, which he wrote about in his book *The Crime of Punishment.* I was especially interested in this quote, as for nearly a decade I worked as a teacher and therapist in the field of primary stress prevention and reduction, drawing heavily on the work of

Dr. Elmer Green and his associates at the Menninger Foundation. Karl Menninger is deceased, so, to get a better understanding of his views, I spoke with a number of people associated with the Menninger Foundation in Topeka, Kansas, including library, archives, and research associates and members of the Menninger family. A reference-library search of Dr. Menninger's writings and a similar search of the archives, including such items as testimony before governmental committees and speeches, revealed only one published article in which Menninger referred to hunting as sadism. In the 1951 article "Totemic Aspects of Contemporary Attitudes Toward Animals" published in *Psychoanalysis and Culture* edited by Wilbur and Muensterberger, Dr. Menninger wrote:

> Sadism may take a socially accepted form. . . . I have in mind, for example, grouse shooting, fox hunting, duck hunting, deer stalking . . . and other varieties of so-called sport. While not quite identical with the horse-whipping pattern, . . . these all represent the destructive and cruel energies of man directed toward the pursuit of more helpless creatures.[8]

Earlier, in Regenstein's book, he states that Dr. Menninger had once been a hunter but gave it up, explaining his reasons for doing so in a letter (undated) to author Carl Bakal, which stated that Menninger quit hunting "after watching a 'few animals die after heroic efforts to escape my long distance bludgeoning. I decided the fun I was having making animals gasp and quiver wasn't worth the suffering it was costing them.'"[9] This statement seems to indicate that Menninger was not a good shot or that he took long shots that he shouldn't have, or that he felt that at one time he himself enjoyed "erotic sadistic" pleasure from hunting.

Today, it is unprofessional to make any generalization about people being psychopathological as a group or as individuals. Our literature search found no published research by Dr. Menninger about hunting motivation or evidence that he conducted such research. In talking with people who knew him, I learned that he frequently wrote impassioned letters about his feelings, and sometimes later had to qualify his remarks or even retract them. What did Karl Menninger really think about hunting?

In *Sparks*, a 1973 collection of Dr. Menninger's writings edited by Lucy Freeman, Freeman states that Menninger was a conservationist who became concerned about certain types of sport hunting, such as shooting polar bears from airplanes, killing hawks and other birds of prey, using strychnine to poison coyotes, and recreational shooting of mourning doves, for which he had a particular fondness. He disliked what he felt was animal exploitation by classic "slob hunters" and worried about extinction of species, but he admitted to liking to eat chicken and turkey, preferring them to dove.

Having said this, Menninger went on to describe how Freud had shown that killing was not madness but perfectly healthy. In Menninger's own words:

> Prior to Sigmund Freud, it was customary in psychology to assume that the wish to kill was pathological. . . . But Freud fearlessly explored the unconscious layers of the personality, and disclosed the fact that it is no more abnormal for a human being to want to kill than it is for a cat to want to kill a mouse or a fox to kill a rabbit.[10]

In *Sparks*, Dr. Menninger talked about how "some" hunters "sometimes" could kill animals by displacing negative emotions and instinctual energies from other settings and projecting them onto animals, as a substitute for their wives, bosses, and so on. He was very careful to not say *all* hunters or *all* the time. He also went on to point out that as people became civilized, they tended to lose opportunities to displace such energies, which could ultimately lead to wars and violence. He acknowledged that hunting, as well as other sports, can serve as an important catharsis of such negative energies.

Sigmund Freud helped us understand how people project sexual themes onto the world around them. A gun can be just a gun,

but it can also be a phallic symbol, depending on who sees it or uses it. A gun is also a symbol of personal potency, and a good marksman would be a potent person, which would mean he had high self-esteem. On the other hand, someone with an inferiority complex could want to own and use a gun to feel more powerful. Like any other symbol, one can project many different images onto it, and these images can be reflections of unconscious issues as well as conscious ones. Someone who fears guns, for example, could have healthy reasons for wanting to avoid guns, or their fears could be a deeper phobic reaction to conflicts about sexuality and repressed anger. Phobias are often triggered by symbolic objects or situations that may not be related to previous traumatic experiences with those specific objects. Guns may be used to intimidate people, but as long as wild animals have the freedom to move about, hunting is more a matter of synchronicity than intimidation. Hunters often return home empty-handed, having fired not even one shot, and yet they report feeling happy and recreated by being outdoors. Someone moved to hunt by sadistic motives would desire not just to kill animals but to see them suffer as a result of their wounds.

In what many experts see as one of the most important and complete studies of violence and aggression ever done, *The Anatomy of Human Destructiveness*, psychiatrist Erich Fromm concluded that hunting was not a sadist pursuit at all, and in fact among cultures around the world where hunting was a primary activity, people tended to be peaceful. This conclusion is supported by many anthropologists, including Colin Turnbull, and leads one to ask whether modern society might be a less violent one if more people hunted, providing they had proper training. As Fromm said:

> Sadism is the passion to have absolute control over another living being, whether an animal, a child, a man or a woman. . . .

The idea that hunting produces pleasure in torture is an unsubstantiated and most implausible statement. Hunters as a rule do not enjoy the suffering of an animal, and in fact a sadist who enjoys hunting would make a poor hunter.[11]

Here again, you cannot say that there are no sadistic hunters. There will always be deviants in any large population of people, but if you look at the psychology of poachers and slob hunters, their behavior is better explained by criminal psychology than by the psychology of hunting. The only place where Fromm found any significant correlation between sadism and hunting was among "elite hunters"—nobility and very rich people who have never grown up and live in a fantasy world protected by their money—who may have high needs for power and control that include a certain amount of sadism. In the late 1800s, King Edward VII and nine companions went shooting in Hungary and reportedly bagged twenty thousand partridges in ten days. This is greed, to be sure, but would not be considered sadism unless these aristocrats purposely made the birds suffer.

Cock fighting, dog fighting, staged fights between bears and dogs, or lions versus tigers, on the other hand, is clearly sadism. Fox hunting with hounds may take on a sadistic nature if the dogs kill the fox. Fromm's conclusion was that the psychology of power elite hunting was essentially a "feudal psychology," and had little or nothing to do with an honest psychology of hunting for either of what he called the "primitive profession" or the "modern passionate" hunters.

Sadism is an act of torture, when the prolonged suffering of a person or animal gives a person pleasure, and the more suffering the better. The 1993 award-winning movie *Schindler's List* by Steven Spielberg is a chilling portrayal of sadism in the Nazi death camps during World War II. In the movie *White Hunter, Black Heart*, Clint Eastwood portrays film director John Wilson, who is

obsessed with the desire to shoot an elephant. Yet, when the elephant is finally cornered, Wilson cannot shoot it. This is an excellent study of immaturity and impotency and their expression in hunting, but it is *not* a study of sadism toward animals.

Moralists often do not care as much about animals as about the behavioral standards of other people, Stephen Kellert's research concluded. Many in his study were more concerned with supporting the exploited and downtrodden and saw poor treatment of animals as just one more case of man's inhumanity. It would not be correct to generalize about moralists, but psychologist Donald Michael has pointed out that some people project their unconscious issues onto social causes, resulting in a "true believer" complex. In such cases, the social cause represents an opportunity for catharsis of unconscious conflict, which in therapy would often be found to be linked to old feelings about Mommy and Daddy, rejections by lovers, or early childhood traumas. Some of the negative animal-rights sentiments may also be the displacement of anger and hatred projected onto hunters as symbols of authority figures. (I have seen individual cases where this was true, but cannot draw any conclusions about how widespread this is.)

Dr. Menninger wrote many books, including *Love Against Hate*, and it seems hard to believe that he would want to see his words used to fuel hatred against others. It also seems unlikely that he would view the long line of U.S. presidents who hunted, including Clinton, Bush, Carter, Eisenhower, Teddy Roosevelt, and Lincoln, as sadists.

In the era when Karl Menninger practiced psychiatry, the goal of many therapists was to help a person adjust to the norms of society. For many today, the goal of the therapy patient, as our understanding of human nature has grown, is self-discovery, which will lead to self-actualization. The goal of many antihunting moralists is to get people to conform to antihunting standards, regardless of whether these standards are compatible with the inner nature of other people. The antiabortion protester who throws blood on a pregnant woman wanting an abortion, sets fire to an abortion clinic, or shoots an abortion doctor is a kindred spirit to the antihunter protestor who slashes tires, burns a taxidermy shop, poisons a hunting dog, or even shoots and kills a hunter—and all in the name of preventing cruelty and suffering! Psychologically speaking, freedom of choice to be who you are and to follow the guidance of your conscience is the most humane ethical position for conservation of the human soul.

What Karl Menninger did oppose was "killing for pleasure," when there was no reverence shown for the lives of the animals killed in hunting, nor use of the animals afterward. He says this quite clearly in *A Psychiatrist's World*.[11] In his extremely popular book *The Crime of Punishment*, which pleads for rehabilitation programs for criminals, Menninger endorses active sports and games "such as skiing, boxing, hunting, tennis playing, and others" as a way to "dissipate aggression."[13] Then in *The Vital Balance* he states that "a small minority of adolescents and adult males continues to find the pursuit of and violent destruction of wild animals peculiarly thrilling." He goes on to say that such "hunting exploits the pleasure of causing fear, suffering, and death in other living creatures, who have as much right to live as we unless their death furthers human welfare in some respect other than amusement." Importantly, he then states clearly, "However, to call such individuals sadistic, uncivilized, and psychopathic is just as untenable as to call their opponents idealistic, fanatic, and sentimental."[14]

Karl Menninger advocated Albert Schweitzer's "reverence for life" and tried to find ways to keep people from hurting one another. The bottom line here is that an accurate "psychology of hunting" cannot be derived from quotes from letters, and anecdotes, and the claim that Menninger felt hunting was "erotic sadism" is not backed up by his published works, let alone any other research. Dr. Menninger clearly said that he did not support stigmatizing people. Using his name to do so is not in keeping with the spirit of Menninger's philosophies at all.

Antihunting accusation: According to famous naturalist Joseph Wood Krutch: "Killing for 'sport' is the perfect type of evil for which metaphysicians have sometimes sought. . . . The killer for sport. . . prefers death to life, darkness to light. He gets nothing except the satisfaction of saying, 'Something which wanted to live is dead.'"[5]

Hunting response: Joseph Wood Krutch was an English professor and drama critic who loved nature. His personal ethics may have been expressed by these remarks, but they are not accurate statements about the psychology of evil or the psychology of hunting. Ethical evil can be anything that is against current ethical norms of one or more people. Sex during the Victorian era, for example, was evil. The same was true for the Shaker community in the United States. During the 1960s, in the Haight-Ashbury district of San Francisco, sexual orgies were common and no one in that subculture saw them as evil. During the same era, hallucinogenic drugs were normally used by many people, while in other parts of the country they were, and still are, seen as evil. People in traditional cultures who use certain psychoactive drugs for spiritual reasons believe that recreational drug use is evil.

Evil in a religious sense is tied to sin, or to digressions from what is seen as the word and laws of God. Religions vary considerably in their evaluation of hunting as good or evil. A Jain or a Hindu might see hunting as evil because it violates certain religious customs and beliefs, but in nearly all traditional cultures, as well as Christianity, Judaism, or Islam, the act of killing animals for food or even religious purposes is seen as normal and desirable.

Sharing this concept with friends in the clergy, I discovered that the Christian Church in many parts of the United States makes a point to honor hunting and hunters. Father Roger Wharton, an Episcopal priest who has had parishes in the Midwest and Alaska, told me that many Roman Catholic and Episcopal churches hold a Hunters' Mass around the opening of deer season. In one of his sermons on the eve of deer season, Father Wharton advised hunters:

This group of people are luckier than the general population as for the next couple of days you will be spending large amounts of your time in the great outdoor world of God's. As you sit in your blind tomorrrow and the next days, or as you walk through the woods—use this opportunity to talk to and listen to your God and King.

Wharton told me that he was familiar with parishes in which the priests asked all the hunters to bring their guns into the church to have them blessed prior to the hunt. He also said that following hunting season many Christian churches hold special wild game dinners to honor the hunters. A hunting buddy of mine in Michigan tells me he and his friends use special permits to take "nuisance deer," which harm farmers' crops, to collect venison for special fund-raising dinners to support his church.

Psychological evil is a different matter from ethical evil. The former refers to an inner psychological process whose motivational roots reside in fear, hatred, and revenge, which overcome normal ego functioning. It is a product of loneliness, despair, and weak self-esteem. Abraham Maslow's exhaustive research on depth psychology led him to conclude that evil was not inherent in humans but the result of the perversion and frustration of needs. If hunting is evil, then you would have to say that the history of the human species is that of an evil animal, and psychological research does not support such a conclusion.

Sigmund Freud believed that there were two primary driving forces in life—"the pleasure principle" of Eros, which moves us to seek happiness and pleasure, and a "death instinct," which balances Eros and is channeled into wars and self-defense or even violent sports. Among Freud's major contributions was to show us how ungratified instinctual drives and repressed emotions cause mental and physical dis-ease. Psychologically speaking, evil is created from unhappiness and fear projected into aggression. Hunters who do not have a psychopathological character, whether modern nature hunters or traditional subsistence hunters, tend to be very happy people; people who are satisfied with life and take pleasure from what they do. Evil, psychological or spiritual, does not occur in self-satisfied people. Hunters do not prefer death to life; rather, they simply seek to participate in the reality of life and embrace it, deriving joy from accepting life as it is. Honest spiritual practices should help a person come to a better appreciation of life as it is and the wonderment of the world we live in. As mythologist Joseph Campbell liked to remind us, "The inhabiting spirit of life is awe." A hunter feels the awe of life from personal experience. If hunting bred evil, then modern sport-hunting organizations would encourage the destruction of all animals, when in reality they have been the leaders of wildlife conservation for decades because hunting has helped them come to know the meaning of Schweitzer's concept of "reverence for life."

Antihunting accusation: Humans have evolved to the point that we don't need to cause animal suffering for our clothes or food. (Source: *The Animal Rights Handbook*, p. 84)

Hunting response: Maybe some humans have evolved, but the vast majority of behavioral scientists worldwide believe that there has not been sufficient time for modern people to make significant changes in mind or body from our Paleolithic ancestors, and that in fact trying to deviate from our instinctual nature is a primary cause of mental disease. Some people can eat a vegetarian diet and stay healthy, if they carefully watch what they eat and use dietary supplements. Vegetarianism, however, which can also be a symptom of anorexia nervosa, may be associated with self-inflicted punishment for guilt feelings about harm caused to other people or animals with a desire to be free of facing the material world. Hitler, for example, stopped eating meat after the suicide of his half-niece, whom he had kept locked in a cage and forced to be his mistress. Charles Manson and his "family" were vegetarians. Such behavior is called a "reaction formation," which means that a person develops behavior that is exactly the opposite of his inner reality so as to deny to himself and others the existence of the underlying emotions and drives.

Carleton Coon, one of the most respected anthropologists of our times, has proposed that we have not yet had time to adapt from our hunting and gathering nature to the agricultural frame of mind, and the dis-ease from the move from being hunters and gatherers to becoming an agricultural society may be a cause of violence and tension in human society. He says:

> Agriculture brings with it a whole new system of human relationships that offer no easily understood advantages, and disturbs an age-old balance between man and nature and among people who live together.[16]

Whenever we deny our instincts, we create problems for ourselves, those around us, and the world. In our inner nature we are all animals. The symbols of our dreams constantly affirm this. As long as our psyches do not change, we will never be able to give up our hunting heritage. The hunting instinct is bred into the bones and blood of at least most of us and is one of the most fundamental elements of human nature. Our challenge as humans is to find the best ways to express our instinctual nature. That is where ethics, values, mythology, the higher self and spirituality come into play as guides enabling us to be healthy, happy human beings.

The sound of sirens is now audible, so it appears that the hunter disrupters will soon be arrested and removed and the hunt can resume. In this last few moments of standoff, the "meow" of a cat is heard, and a house cat emerges from the bushes. This now gives the hunter a chance for a parting shot.

> One of the big problems which we've got these days is wildlife predation by domestic cats. A recent study by the University of Wisconsin found that farm cats kill at least 15 million cottontail rabbits and 28 million songbirds every year—in Wisconsin! They kill this many because there are about 1.6 million farm cats in Wisconsin.

Multiply those figures nationwide and you'll soon see that house cats kill far more small game animals than all the hunters, and in many cases more than many of the natural predators. If you want to help cut down on suffering let's talk about what we can both do to drastically reduce the domestic cat population that is semi-wild or completely feral.[17]

The arrests have been made and the hunters now return to the blind. As they settle in, someone notices that a few of the decoys look new. At that point, the new "decoys" take off, before anyone can raise a gun, causing spirits to rise again. The morning flight is not over, but it will take some time to get settled back into the spirit of the hunt.

Instinctually, we are hunters, and have been since the days of Olduvai Gorge 4 to 5 million years ago. Current archaeological thinking is that in those days there were two primary species of biped man-apes, a vegetarian, *Australopithecus afarensis*, and an omnivore, *Homo habilis*. The evidence suggests that as man evolved into *Homo sapiens neanderthalensis* (Neanderthal man) about 400,000 years ago, and then into *Homo sapiens sapiens* (modern man) about 50,000 to 30,000 years ago, vegetarianism remained with gorillas and most monkeys, while other species became omnivores by nature. Chimpanzees, our closest ape relatives, are omnivores, and Jane Goodall found that they go on hunting parties about thirty times a year, sometimes using sticks and stones as weapons.

If we agree with Freud and William James that hunting is instinctual in man, at least in some of us, then the urge to hunt that was known to our ancestors at Olduvai Gorge still must live within us. Like any instinct, the desire to hunt can be a powerful human motivation. We cannot return to the past, at least not in any numbers. In general, hunting and gathering cultures have population densities of ten square miles per person. Yet, the same powerful instincts are still there, and denying them always has its costs. The modern hunter, therefore, faces two challenges unknown to his predecessors: how to keep in touch with the hunting instinct and keep the embers of its fire from going out amid the necessary constraints of living in modern times; and how to manage the hunting instinct in a way that is in keeping with contemporary cultural norms and laws. Now, more than ever before, hunters require guidance to carry on the hunting heritage.

Guidance for the Hunt

Before ethics were created, hunters divined, consulting the gods for guidance on locating game and how to show proper respect for what was killed. Hunger was the motivation for early hunting, and all around the world the vast majority of humans hunted to satisfy their omnivorous needs. Typically, shamans or priests would make offerings and consult spirits through trance, dreams, or reading omens. In some cultures such divining took place only at special times of the year. Among the Nganasan, a surviving tribe of the Taymyr Peninsula in northern Siberia, a shaman would make spirit journeys to the seven spirits who govern hunting success and community well-being only on three days of the year: the winter "Festival of the Clean Tent"; "The Great Day" of the spring, usually around June 20; and "The First Snow" in autumn.[18]

Traditional hunters pray to certain spirits or gods who preside over hunting, such as the Eskimo's Old Woman who lives at the bottom of the sea, Ogun and Ochosi in West Africa, the Mother of the Animals, Artemis or Diana, and songs, stories, legends, and rituals about these supernatural beings preserve the ethics of hunting. Among traditional cultures around the world, then and now, the spiritual and the material dimensions of life are brought together so that awe and wonder may be everywhere. Myths, to the traditional hunter, are not just stories recorded in books, they are alive.

In many cultures the guardian spirit of the hunt was female. In the Finnish epic poem the *Kalevala*, before going off to hunt elk, Lemminkainen invoked spiritual forces with charms and prayers, making a special plea to the daughter of Tapio, king of the forest:

Girl of the forest, lovely maid, Wind Spirit, Tapio's
    daughter!
Drive the game to the sides of the way, to the most exten-
    sive clearings . . .
Make it run fast, dash swiftly in front of the man in search
    of it.

In ancient Greece and Rome, a chaste woman, Artemis or
Diana, who was the "Mother of the Animals" and presided over
birth and death, was the guardian of the hunt. Artemis was also
said to appear in the form of a bear, or a bear-woman, and among
many cultures the bear is considered the most powerful and wisest of all the animals. Failure to heed the guidance of the gods not
only brought on failure in hunting but could result in bad luck
in general and in illness. Among the Lakota tribe in America,
their counterpart to the Christian Jesus was White Buffalo Calf
Woman, who could change from the form of a woman to a white
buffalo, which was their principal source of food and clothing.
Later we will see that the number of women who hunt today is
growing at a surprising rate.

Moses was an early wildlife biologist, decreeing that when a
bird's nest with young birds in it was found, the young should not
be taken, but in general, modern religions do not have a lot to say
about hunting. The Christian religion does not disapprove of
hunting, and Genesis 1:28 counsels us to "Be fruitful and multiply,
and fill the earth and subdue it; and have dominion over the fish
of the sea, over the birds of the air, and over every living thing that
moves on earth." The New Catholic Encyclopedia, in addition, advises, "Unnecessary cruelty must be avoided, such as making captured animals suffer for a long time. For this is an abuse of a
creature of God and is degrading to those who practice it." Jews
who follow kosher traditions are forbidden from eating the flesh
of pigs or fish that have no scales, but there are no prohibitions
against hunting. Muslims eat meat, and some of their early ceremonies involved animal sacrifice. Their only prohibition about

eating the flesh of mammals is that they should not take meat from an animal that is already dead, which seemingly encourages hunting. Buddhists can eat meat, but many follow the ahimsa practice of not killing animals. Cookbooks of the peaceful Amish contain many tasty recipes for meat dishes. Among the major world religions today, only the Hindus and the Jains forbid hunting and eating meat. Manichaeanism, a religion popular in the Middle East fifteen hundred years ago, advocated eating only fruit, but little is heard from Mani's followers today, and I have my suspicions why.

In Asia, Marco Polo reported that Kublai Khan was a great hunter, and also a conservationist. Polo reported:

> There is an order which prohibits every person throughout all the countries subject to the Great Khan from daring to kill hares, roebucks, fallow deer, stags or other animals of that kind, or any large birds, between the months of March and October. This is that they may increase and multiply; and as the breadth of this order is attended by punishment, game of every description increases prodigiously.[19]

In more modern times, with the separation of church and state, the dominance of science, and the development of complex urbanized lifestyles, game laws—rather than myths and gods—increasingly have come to regulate the limits, seasons, and methods of lawful hunting, especially where ownership of game is public, and especially for migratory waterfowl that move across international political boundaries in their annual migrations. The first game law in the United States established by Europeans was the 1646 law in Rhode Island to protect deer. In 1699, Virginia forbade killing deer between January and July, with a penalty of five hundred pounds of tobacco for violators. Today, all states have very detailed regulations on hunting, sometimes almost smothering it because of their complexity and detail. In one recent study of Michigan hunters who left the sport, the primary reason was feeling overwhelmed by the growing number of regulations. The detail of modern game laws is due partially to modern scientific methods of wildlife management and hunting and fishing restrictions on some species, but it is also a reflection of the defensiveness of many fish and game departments when faced with challenges by animal rights groups.

Beyond laws, there are additional factors that today guide hunting, setting codes for behavior in excess of what other predators or subsistence hunters follow over and above what the game laws spell out. "Sport" hunting may be an unfortunate choice of word to describe modern nonsubsistence hunting, for the goal of the true sportsman of the blood sports is to give the game the greatest chance possible to escape, and then, if caught, to kill the animal as quickly and humanely as possible. In 1893, Theodore Roosevelt put into words the concept of Fair Chase, as a sport hunting ethic to limit the manner in which big game trophies could be taken to qualify for recognition in the Boone and Crocket Club records. In the early Boone and Crocket Club publication "American Big Game Hunting," Roosevelt set out the ethic:

> The term 'Fair Chase' shall not be held to include killing
> bear, wolf or cougar in traps, nor 'fire hunting' nor 'crust-
> ing' moose, elk or deer in deep snow, nor killing game from
> a boat while it is swimming in the water, nor killing deer by
> any other method than fair stalking and still hunting.

Later on, Fair Chase was broadened to forbid herding animals by air or any motor-driven vehicles, the use of electronic communication devices for locating, attracting, or observing animals, and hunting game confined by escape-proof fencing. Today, these practices are still illegal, as well as unethical, in most states and countries.

Aside from recognizing how modern sport hunting seeks to handicap the hunter and give the animals maximum chance to escape, it should be noted that in earlier times, traditional hunters often hunted by driving game with fires, and preferred to hunt in deep snow or by chasing animals into water where they could be more easily killed. Herding animals were frequently driven over cliffs or into pits. This was utilitarian hunting, not sport hunting, but people did derive enjoyment from it and later may have held celebrations to honor the success of the hunt and the hunters. Traditional subsistence hunters are practical in their methods, trying to get the best results possible, for their success may mean life or death for themselves or the entire community. There was probably much more suffering among animals taken by traditional methods than those taken by modern hunters.

An ethic is a philosophy of morals, a standard for character for a group or race. The word *ethic* originates from the Greek word *ethikos*, which is the characteristic spirit of a people. In contrast, *pathos* is the quality or element in speech, events, art, or action that arouses emotions or passions. Where the animal rights arguments fall apart, it seems to me, is when pathos becomes more a part of the statement than ethos.

Probably the most practical and clearly stated ethic for hunting, and for judging environmental action in general, was proposed by Aldo Leopold in *A Sand County Almanac*. Leopold's "land ethic" is:

> A thing is right when it tends to preserve the integrity, stability, and beauty of the biotic community. It is wrong when it tends to do otherwise.[20]

Hunters take pride in being successful, for after all, if you would be satisfied with shooting pictures, you would not carry weapons into the field. Concerned with a quick, clean kill, every hunter needs to learn the range of his weapons. But there is satisfaction in the skill of hunting as well as the kill. There is an element of challenge in the longer shot. Success proves your marksmanship, and making a longer shot is something you will remember. In Europe, where hunting opportunities are very limited, making the difficult shot is stressed more than trying to fill one's limit.

If I am using a rifle and the distance between me the animal is considerable, I always aim directly for the head or heart and wait for the time when I can make only a clean hit or a clean miss. That way, if I make the shot, the kill will be instant. I clearly recall shooting a rabbit with a .22 rifle at twilight at a distance of at least one hundred yards. I aimed for the head and hit him exactly where I had aimed, without a telescopic sight. Thirty-five years later, friends still talk about that shot and the rabbit stew we had that night for dinner.

Sights are of little value to the shotgunner, for the targets are usually moving too quickly to sight in on anything. You "point" a shotgun as a flowing movement where the wisdom of your eye and your sense of distance help put the muzzle at the right place when you pull the trigger. Much the same is true for the instinctive archer. You must aim by becoming one with your weapon and then direct your intention toward the target. The Zen archers maintain that you ultimately shoot yourself by blending your consciousness with that of your target.

Some hunters attempt to shoot the largest animals available for trophies, seeking to get their names in the Boone and Crocket Club registry. An integral part of interior decorating for many people and places are heads, hides, horns, feathers, and fins carefully sculpted by taxidermists into lifelike representations of the real thing. These are memories, works of art whose presence recalls the time when the life of an animal was taken. Animal rights people tend to jump quickly on trophy hunting as *always* being an ego trip, and taxidermists have sometimes been the targets of vandalism and death threats. Several very good taxidermists in my area do no public advertising for fear that their homes and families will be harmed.

Psychologically, there is a big difference between needing *always* to kill the largest or biggest animal and happening to kill an unusually large specimen. The former could be someone with immaturity problems, the latter may simply be a blessing and testimony to superior skill. In this day of modern wildlife science, it is possible to manage lands and animals to create more trophy-size animals. You need the right food, the optimum ratio of males to females, and satisfactory habitat. The thought of this approach to wildlife management makes many animal rights activists howl and yet how this differs from domesticated animal breeding escapes me. There is nothing inherently wrong with mounting a deer's head and hanging it on the wall. For some people it could be an ego trip, but for others it is the memory of a lifetime, and the presence of the deer's head is a constant reminder of the need for conservation. In some cultures, the same head could even be a religious icon.

Among the Huichol of Mexico, if you are on a hunt for the special deer whose blood and meat will be used for the Ceremony of the Drums to honor the children of the village, the hunter does not shoot until the deer looks at him, their eyes meet, and a sense of mutual understanding is felt. In traditional arctic hunting tribes, like the Nganasan, where meat is a primary source of food, the hunters may prefer to take the largest, healthiest animals, such as the biggest caribou or reindeer. Such animals take the most food, and if they are killed, there will be more for the younger animals to eat during the winter.

All the seasoned hunters I have ever known speak with quiet reverence of knowing which animal or bird to shoot. It is not something that may be felt right away. Like any practice, it takes time, but it is a sense that has been the consistent guide of master hunters since the Paleolithic.

In the late 1970s professors Robert Jackson and Robert Norton of the University of Wisconsin, LaCrosse, interviewed more than one thousand hunters to study the development of the hunter's motivations.[21] As hunters start out learning to hunt, which Jackson and Norton called the Shooter Stage, they are most concerned with having some success to demonstrate their competence. The first kill seems guided by a mixture of innocence and eagerness for many. Realizing that you have taken the life of another living thing may or may not be at the forefront of your awareness if you just shot his or her first deer or goose. You are just beginning to learn to use a weapon and hunting is still very new to you.

The second stage they call Limiting Out, and here the goal becomes to take as many animals as the law allows. Hunting success indicates just how well a hunter has mastered the required skills.

Having demonstrated that it is possible to take a limit, hunters begin to be more selective about what they shoot. Jackson and Norton call this the Trophy Stage. Along the marshes of Lake Erie in the 1950s this meant taking only black ducks or drake canvasbacks, as opposed to any duck you could shoot. The theme of this stage of hunting is that the hunter has gained enough proficiency to know that he or she can kill something if needed, but now a little more challenge is added to keep the excitement levels high. Some people accuse trophy hunters of taking the largest and most vital members of a population, thereby weakening the gene pool. In limited populations, this would be possible, but in general, Darwin's survival of the fittest seems to result in animals getting more and more sophisticated, as well as larger. In refuges across America, flocks of ducks are seen daily in protected areas feeding peacefully in range of hunters in blinds located on range open to hunting. In heavily hunted areas of northern California, such as the Lake Sonoma Recreation Area south of Cloverdale, there are massive feral pigs and wild boar and many big black-tailed bucks, despite intensive organized hunts that have been going on for years. The solution to the dangers of trophy hunting weakening a population is to have enough information about those animals to regulate the number and kinds killed. The truly wise animals are those that continue to survive heavy hunting.

Not all selective hunters, however, try to take the biggest species. Some are moved by humane motives to shoot crippled animals, or the oldest ones, and ones that would likely die of starvation, instead of others in the prime of life. This is exactly what other predators would do. Ethics at this stage are personal and may have absolutely nothing to do with needing to prove anything to anyone. It is really what role you see yourself playing in the ecosystem and what the values are that move your life that determine which animals you will shoot and which you will let go.

Next is the Method Stage, in which the hunter places even more limitations on the methods of take, such as using a muzzle-loading rifle or shotgun, a bow and arrow, a slingshot, or even a modern version of the atlatl spear thrower. The hunter moves closer and closer to the animals and takes more and more care to make a quick, clean kill when the target is chosen. With growing fondness for game animals and nature in general, hunters at this stage also frequently take along a camera and increasingly find pleasure in nonconsumptive wildlife watching, as well as hunting, for they are interested not in limiting out or taking trophies but rather in maximizing the intensity of the experience, calling all their senses into play in a fashion not unlike that of their ancestors.

Last comes the Sportsman Stage. Over the years there is a deepening of the experience of hunting that naturally leads one into a mental set where the most poignant and humane of values come to mind when seeking guidance about what to shoot—ecological balance, harmony, reverence for life, stewardship, even love. Hunters at this stage may turn to art to express their fondness for hunting and animals and often devote considerable time and money to helping preserve habitat and teaching younger people to hunt. This is the golden age of hunting.

As might be expected, as time passes, the number of hunters declines, according to this research. This is to be expected, as people experiment with a sport for many reasons and stay with it or drop out according to their satisfaction.

Thinking back to hunters I have known, looking at elements of the progression of my approach to hunting, and comparing these with Jackson and Norton's sequence, I think their five stages of hunting may be descriptive of hunters in general, but since the completion of their research there has been increasing evidence of a growing number of what Stephen Kellert calls Nature Hunters, who enter hunting as an almost spiritual pursuit from the beginning. Whereas the average hunter is in his early forties, nature hunters tend to be younger and have at least some college education, many have advanced degrees, and many have studied native cultures or the martial arts, and all have a keen interest in nature that is maintained year-round. They enter hunting as a sort of personal calling, perhaps without having been taught by parents, and they take the sport very seriously. These are not people who trespass, take "sound shots" in the woods, or skyblast. They seem more akin to the rapidly growing number of fisherman who go fly fishing for trout, as they choose to use more "primitive" weapons like archery and muzzleloading rifles and shotguns. As I'll discuss in a later chapter, their numbers are rapidly growing while hunters in general may be declining.

A stuffed elk head on a wall may be the symbol of someone who is immature and has an inferiority complex and a tremendous ego need to compensate by conspicuous displaying of trophies. The same stuffed animal can also be seen as an icon that states that so-and-so has a unique kinship with a species that has resulted in hunting success, and that his values are shared with traditional hunters who often mount skulls, horns, hooves, and other parts of animals taken in prominent places as a way to honor the spirits of the deceased. Once I was given the special blessing of a world-renowned Eastern spiritual teacher of being invited into his private meditation chamber, which only a handful of people had ever seen. I was expecting to see the room filled with statues of Buddha, which were everywhere else around the house. Instead, what I found, to my great surprise, was the stuffed head of a stag on the wall above his special chair, and beside him a full-body mount of a lion. He explained simply that these were his guides in meditation, helping him keep in sympathy with their wisdom, which was the spiritual intelligence of nature. Similarly, I have known yogis from India who were vegetarians who kept bobcats for pets, and in earlier times they would have kept lions, leopards, or cheetahs.

In time, a true spiritual seeker goes through a period called "the dark night of the soul," when the shadow side of the person is brought out. The one who emerges from the other side of this ordeal, and has integrated both his darkness and lightness, is a person of power. Carl Jung called such a person a "mana person," and said that such people are living archetypes who have gained mastery over their whole self.

Hunting, or shooting as it is called in England, can be done for many different motives, positive and negative. It is also possible that some animal rights activists have taken up their cause because they are emotionally immature and are acting out repressed rage and hatred from emotional wounds from parental abuse or from failed love affairs, seeing the plight of "defenseless animals" as symbolically similar to how they may feel they have been personally treated by a dominating human, who is symbolized as a hunter with a weapon.

You cannot draw conclusions about the motives behind any act without knowing about what is going on inside the mind of the person, as well as the situations under which it occurred. Sex can be love or rape. Killing another person can be self-defense or murder. Vegetarianism can be an act of spiritual devotion and greater health, and it can also be an early symptom of guilt-driven suicide. There are hunters who kill out of season, take more than their limit, skyblast, trespass, drink while hunting, shoot carelessly, and leave wounded animals in the woods. Rather than blame all hunters, animal rights people ought to be working together with ethical hunters to get these bad eggs out of hunting. Today, more than ever, the hunting community needs to come together and seek to make hunting a very ethical sport. This is the surest way to secure hunting privileges on public land for the future, as public scrutiny intensifies and is not likely to diminish in the future.

I'll tell you what my hunting ethics are, in addition to obeying game laws. First is to not shoot unless I am reasonably sure that I will make a quick kill. Second is not to shoot anything that I will not eat. Third is to track down every animal wounded by myself or my companions, to the extent possible. And fourth is to not let any meat spoil from game I shoot. Beyond this, I have adopted a system of what I call "the three prayers of a hunter." As I first set foot on the land where I want to hunt, I begin with the advice of my Lummi friend Kenny Cooper and say some prayers, affirming these ethics and stating my intention and then I leave a little food as a sacrifice. When I see something that I could shoot, as I am about to shoot, I say a silent prayer that I want to shoot only what is right for me to shoot. After I shoot something, I say a quiet prayer of thanksgiving, which includes the pledge that I will do what I can to make sure that the ancestors of this animal will have the habitat and protection they need to survive. Many times I have passed up sure shots just because they did not feel right. Following these steps doesn't mean that I get more game than anyone else; in fact, it may mean that I get less. But what I do kill I feel good about, and on some occasions, certain animals seem to come to me or I am guided to them when I am hunting.

SOME DUCKS SEE the decoys and I take up my call and make the hailing call. They begin to set their wings to land. Now I change the call to the feeding call. They are getting closer and closer. At forty yards they are in range. They are widgeon—all legal birds. Remember to shoot only at birds that are in front of you, or off to the side, if no one else is at your side. I say my silent prayer to let me shoot what is right and I feel moved toward one bird in the flock. I can't say why, but it feels that this is the one I should concentrate on. I see a second if I am successful with the first, but one at at time. Okay, it's up to you now. Shoot with your gun or camera, or just sit and watch. Do what feels right to you in your heart and soul and the guides of the hunt will always be pleased.

A hunter should never let himself be deluded
by pride or false sense of dominance. It is not
through our power that we take life in nature;
it is through the power of nature that life is
given to us.
RICHARD NELSON[22]

Six

Tools, Talismans, and Weapons

I can still remember the first time I pulled a trigger on a .410 shotgun, because I was too little to hold a 12-gauge. This is part of the culture of a big part of America.

PRESIDENT BILL CLINTON,
after returning from a duck hunt, December 27, 1993[1]

We cannot but pity the boy who has never fired a gun; he is no more humane, while his education has been sadly neglected.

HENRY DAVID THOREAU[2]

"You should never point a gun at anything that you do not intend to shoot; I don't care if it's a toy gun. Treat every gun like it's loaded. If I see you do it, I'll take it away and you'll never see it again."

This was my father's stern rule for me when I was old enough to want a cap gun. When he said it, I protested, because I knew that he meant it. Other kids had cap guns and played with them, shooting at each other in play, I argued.

"You're not other kids," my father said. "You develop habits when you shoot a gun, because sometimes you don't have time to think a lot before you have to shoot. Treating a gun as if it were always loaded is the first habit to develop." My father agreed with Mark Twain, who once said that there is no more dangerous weapon than an "unloaded gun."

Consequently, I never had many toy guns when I was growing up. The only one I recall looked like a double-barreled shotgun that shot corks. I think I was about five when I got it, and I do not remember ever taking it outside to play with other kids. When they played cowboys and Indians, I always ended up an Indian, because Indians had bows and arrows instead of guns.

On the other hand, I got my first real gun at seven. It was a single-shot BB gun, a Ryder, I think. I was allowed to use it only with my father, and we practiced in the basement for quite a while before he let me take it outside.

I brought it with me when my father took me duck hunting, and I shot when he did. It was not really a lethal weapon for a duck, but I was very proud to be like the grown-up hunters.

The first living thing I ever shot was with that gun, a sparrow. My father had told me that sparrows were nuisance birds, and he showed me in the hunting regulations how you could shoot sparrows, starlings, and blackbirds year round, because they damaged farmers' crops. I remember that first kill vividly. I saw the bird, raised the gun to my shoulder, took aim on the bird's head, and squeezed the trigger. The BB went exactly where I had aimed and the bird crumpled. I was awestruck. Prior to that time, paper targets and tin cans were all I had shot. In that moment I realized that I held the power of life and death in my hands when I had a gun. I have never forgotten it.

Kids who watch television and movies see violence all the time; it is distant from them and even entertaining. When you first kill something, see the blood and hold the lifeless animal in your hand, it is a moment you will never forget. If people were responsible for killing even a small portion of the meat they eat, animals in general would be treated with more respect and compassion. Part of the cause of the violence in society today is that people live far too much in fantasy, never developing an emotional awareness of the consequences of their actions (like eating meat) that support death.

When I was about ten, my father gave me a .22 pellet rifle. I was very proud of this gun because it was real, meaning that it was powerful enough to kill the things grown-up hunters shot at, like ducks, rabbits, and pheasant. It also was a perfect gun to shoot rats, and there were sometimes hordes of Norway rats that at dusk would come out from their homes along the banks of Frenchman's Creek and venture up into people's yards. The rats were not native, my father taught me, they had been brought over by accident on ships, and they were carriers of disease and were taking over the habitat of the muskrat, which was a native species. Shooting rats helped preserve native animals that were also valuable for food and skins. Varmints, I learned, were an exception to the rule that you never shoot anything you don't want to eat. My opinion of rats went even lower when I saw that sometimes after I shot a rat, its former friends would come out and eat it.

Fur was still considered fashionable in those days, and during the winters my father trapped muskrat and mink out in the marsh. The water business was slow during the winter, and the pelts and carcasses brought in some welcome extra income. The fur buyer told us the muskrat would become "Hudson Bay seal" and the carcasses "marsh rabbit," as people didn't like to think they were wearing or eating rats.

Gradually I took over the trapline and made enough to buy a boat, a fourteen-foot wooden rowboat that was like a square-stern canoe. I was allowed to go out in the marsh in the boat by myself to hunt with my pellet gun. I think I was eleven. I could not shoot the pellet gun into the air, my father instructed, as the pellet would carry too far. This meant that if I was going to get anything I would have to sneak up on it and shoot it on land or on the water. Learning to pole the boat noiselessly through the marsh was a prerequisite to success. The first wild game I ever shot was a sora rail, which is a relatively small marsh bird about the size of a quail with long spindly legs. I remember well that we had a celebration that night, baking the rail with a strip of bacon around it. The bird was smaller than a game hen, but to me it was a trophy that I will never forget. It was a milestone in life, a milestone of growing up.

Rail are easy to hit with a shotgun when they flush, because they do not fly rapidly, but when they are stalking through the dense rushes and cattails, they are like tiny ghosts. I did not shoot many rail, and it was very difficult to get close to ducks, except those that had been crippled, some by natural causes like flying into power lines at night, and others shot by other hunters but not killed. The wounded birds would hide in the marsh. Normally, duck hunters hunt only in the morning or late afternoon, as this is when the ducks fly best; they would fly out into Lake Erie during the day, where they could not be easily approached, then return to the shallow water at night to feed. With my pellet gun, I would go out during the middle of the day, after the morning shoot, and become a mercy killer of crippled ducks. My chief competition for crippled ducks were foxes, raccoons, skunks, hawks, owls, cats, and the Lake Erie rats, and, believe me, they got far more than I did. In my experience, crippled ducks that cannot fly rarely escape natural predators for very long. Today in some heavily hunted marshes, like Pointe Mouillee near Monroe, Michigan, hawks wait in the trees for wounded birds that hunters cannot locate. The warden at Pointe Mouillee tells me that every winter snowy owls move into the marsh to catch any cripples the hawks don't get.

As soon as the duck season was over I put down my gun but continued chasing ducks. The ones we pursued were those that had been caught in oil spills that came drifting down the river at night, catching thousands of ducks unaware. The oil would mat down their feathers and they would lose their normal downy buoyancy. One coat of oil and they would lose their weather-proofing, and they would be heavier and less tolerant of the cold. Waterfowl would try to preen the oil off their feathers, the way they preened natural oils, but when they swallowed the crude they would be poisoned, on top of being all gooed up. The late

George Hunt, my wildlife management professor and later hunting companion, estimated that ten thousand to twenty thousand ducks and geese perished each winter in the Lower Detroit River from oil spills. Not much was said about this because most victims of oil spills are never seen, Hunt discovered. Their feathers become caked to their bodies, they get poisoned trying to clean themselves, and they simply become too weak to fly, quietly slipping under water and sinking to the bottom. The graphic photos of oil-soaked birds lying on the beach after a spill, like the massive one in Prince William Sound in Alaska in 1989, are just the tip of the iceberg of wildlife casualties from oil spills.

We would chase the oil-soaked ducks in the boat, as they could not fly, and catch them with nets, take them home, and wash them with detergent to get the oil off their feathers. While they were recovering, we would keep them in a pen and feed them cracked corn until they were strong enough to fly away. Those were some of my best school science projects, and the rescuing was a very natural thing to do; a way of saying thank you to relatives of birds I had shot, and at least a small statement of apology for the oil pollution caused by humans.

My first shotgun was a single-shot 20-gauge, but I had already learned to wing shoot with .22 rifle that had a smooth bore and was loaded with tiny bird shot shells. We hunted grasshoppers with that .22, and grasshoppers in flight are about as hard to hit as pheasant or quail, although the distance is shorter. We hunted grasshoppers because they were pests to farmers.

I was about twelve when I got the 20-gauge. I was proud of it, but soon I learned its limitations as it had only one shot and not as much potency as a 12-gauge. Ducks are not always easy to kill, and after I brought one down and it got away because I couldn't reload fast enough before it dived under and escaped, I pleaded for a double-barreled 12-gauge like my father's. My grounds were humane: The better shotgun would enable me to kill more reliably and quickly anything I shot. Soon I got a Winchester double-barreled 12-gauge, which I used until steel shot became available.

The old barrels could not handle the new steel loads, so that model was retired. The graduated approach to learning to use a gun makes the shooter feel increasingly self-confident, able to assume more and more responsibility for weapons that are more and more powerful. One of the most important measures of maturity is how well a person can handle potentially dangerous situations and materials. I welcomed the introduction of steel shot, as spent lead pellets eaten by ducks and geese can cause lead poisoning, but having to get a new gun made me feel sad, like losing an old friend.

Weapons separate hunters from wildlife watchers. Today in America there are at least 200 million guns, almost one for every person, and gun ownership is rapidly growing. Some people enjoy shooting targets, and skeet and trap shooting are growing in popularity; others collect guns as a hobby. But the majority of guns people own are for hunting, self-defense, or breaking the law. And as the number of hunters has been dwindling in recent years, you know who is buying guns. This definite switch in gun ownership patterns speaks to a rising tide of fear and violence in America not at all related to hunting.

It is estimated that 65 percent of the homes in the United States have a gun, which means that gun owners seldom have just one gun. In my experience, when a person owns just one gun, usually it is a handgun for self-defense. In California in 1993, gun sales made records. There is a mandatory two-week waiting period for anyone to buy a gun, during which time his or her records can be checked for history of crime or mental illness and to certify that the purchaser is twenty one or older, so it is possible to know more about gun owners than ever before. Of the 665,229 guns sold in California in 1993, 448,247 were handguns, and you can be sure that most of those were *not* for hunting. In 1993, the California Department of Justice denied sales of guns to 6,509 people, about half of them because they had records of convictions for violent crimes.

Hunters' guns are nearly all long guns, shotguns and rifles. Long guns are not a good choice for criminal use, for they cannot be easily concealed and are too long to be easily maneuvered in tight situations.

People who own guns for self-defense tend to hide them away and not shoot them very much. Hunters' guns are more public. Often they are kept on display in gun cases or on racks on walls in full view.

There is a different feel about a hunting gun. It carries fond memories. Used with skill, it provides special meat for the table, healthy meat, meat that carries the spirit of the wild and has flavors that were known to our ancestors millions of years ago. To eat wild meat is a sacrament for many people. It is eaten at special times and shared with special people. Eating it brings back fond memories and kindles a spirit of thanksgiving as you recognize the wild animals it came from. When I eat the flesh of wild animals, I feel different, more alive. The Huichol shaman Don Jose Matsuwa, who recently passed away at 110-plus years, insisted that the meat of wild animals had more *kupuri* or life force energy, as well as less fat, less cholesterol, and more protein than meat from domesticated animals.

Weapons may carry spiritual values too, not unlike the ritual knife that a kosher butcher, called a *shohet*, uses to kill animals. According to kosher tradition, the *shohet*'s knife, used to cut the throats of the animals, must be razor sharp with absolutely no nicks in the blade. The purpose of using a knife for slaughter is to kill the animal quickly and allow it to bleed to death, for Jewish tradition forbids eating blood. The customs of Jewish slaughter are dictated by *hukim*—religious laws that cannot be fully explained. This is a theme that we will return to later, as weapons need guidlines to make them work on *behalf* of people rather than *against* them. King Arthur's sword, Excalibur, is another case of a metal weapon bearing mysterious powers. It is said to have come directly from a spirit, the Lady of the Lake, who gave it to Merlin the Magician. She told him that whoever used the sword would be the king of England. Merlin plunged the sword into a boulder.

Because only Arthur could extract the sword, he proved his right to the throne.

Guns also have personalities. Hunters will fondly talk about "old Bessie" or "my deer rifle" as though they are dear friends. Sometimes it is hard to say whether the guns take on the personalities of the hunters or the hunters take on the personalities of the guns. The same is true for bows and arrows. The bond between hunters and their weapons runs deep. I believe a hunting weapon is like a lightning rod. It carries the power of life and death, which means it is a tool of great respect, even fear. Mastering the use of a weapon requires self-control, which builds confidence, whether we are talking about a gun, a knife, a bow and arrow, or the lethal hands and feet of a karate master. Weapons are hard and cold, and using them correctly requires self-control and concentration. They are an extension of the people who use them, a magnifier of personal impact and attitudes.

The Freudian interpretation of guns and other weapons as phallic symbols seems obvious. The symbolism of the barrel as a penis is clear, as is the relation of the bullet to semen. Arrows and knives penetrate with force. One of the most famous guns in modern times is Clint Eastwood's .44 magnum, the "Dirty Harry gun" from the 1973 movie *Magnum Force*. Critics called it "an angry erection." The movie made Eastwood's character a legend, establishing an archetype of the American consciousness. As Eastwood says, the only two things that America ever really invented are jazz and western movies, and Dirty Harry is ultimately a modern cowboy.

A weapon also carries history with it. When you use a weapon, its lethality brings you into one of the most basic of all experiences of life—killing. In my gun cabinet there are both guns and bows and arrows. They take on even more personality when I see them as modern evolutions of one of mankind's first technologies. Tracing a brief history of each adds conscious meaning to their unconscious symbolism, enabling a better understanding of their unique qualities.

A Brief History of Hunting Weapons

We can see, in the famous scene from the movie *2001: A Space Odyssey*, when the ape-men first discover that a bone can increase personal power and be used to increase one's ability to kill, the possibility of the discovery of the first weapon. Others suggest that thrown stones and hand-held sticks were first used to hunt and defend against predators long before fire was tamed. Sifting through the relics at Olduvai Gorge in Tanzania, archaeologists tell us that 5 million years ago there was a split in the *Australopithecus* type of biped apes. The *Paranthropus* type, primarily vegetarians, lived in more moist riverine forests, gathering roots, nuts, seeds, fruits, and berries for food, which was better suited for their large, wrinkled, grinding teeth. On the drier uplands lived an omnivorous *Australopithecus*—with teeth more like our own—who ate lizards, snakes, rodents, and the young, aged, or ill members of larger species. They may have acted as scavengers, following the kills of predators, but the staved-in skulls of baboons, antelope, and other small mammals found with the bones of our ancestors indicate that they also killed live animals for food.

The early ape-men began to be tool users as they discovered that a hand-held rock or stick increased their ability to kill. Even today we call weapons "arms," which suggests that we see them as extensions of our bodies. Who knows how long it took for early man to discover that it was possible to extend the range of their ability to kill by hurling stones or sticks.

Homo habilis, a descendant of *Australopithecus*, was probably the first tool maker. His cranium was larger and his hands had the humanoid grasping ability, both of which made hand-held tools a natural extension of self. A tool user picks up a stick and throws it. A tool maker sharpens the point of the stick first.

All modern weapons are descended from two basic early types of weapons—the pebble became the bullet, and the stick became the spear and then the arrow. Tracing the history of each yields insights into their spirits.

Sticks to Compound Bows and Rock and Roll

Sticks and bones were among the first human tools. They extended the reach and enabled an increase in the power of the human arm, enabling important advances in digging, hunting, and warfare. A stick can be used as a potent weapon, as is seen today in the fighting jo-staff used in many of the Asian martial arts. The striking distance of the stick increases when a point is carved on its end, making a lance or, when thrown, a spear, and later a javelin, capable of being thrown accurately over much larger distances.

Adding a special head of stone or bone combines the potency of a knife, with its cutting edges, with the added distance of a spear. Success or failure with a spear meant life or death, and among hunting and gathering peoples it was common to carve or paint images on the shaft or head. This unique pattern served to identify the weapon, making it easy to tell who had struck the lethal blow in the heat of the hunt or battle.

In earlier times, in many cultures, it was believed that weapons had a soul or spirit, as did the owner of the weapon. Invoking the spirit of the weapon was thought to increase its accuracy. The Inuit, for example, believe that tools such as spears have an *inua* that can work for or against the hunter. In making the tool or weapon, ceremonies and rituals frequently were conducted to invoke the spirit of the tool, asking for its cooperation so that the tool would become an ally.

In many cases these ritual sentiments were translated into art through decorating the shaft of the spear with carvings or paintings or attaching items such as feathers, skin, and bones. The artwork transformed the utilitarian tool into a numinous object, for in traditional cultures art is an invocation to powers beyond the material plane. Magic is based on two principles, sympathy and contagion. The principle of sympathy is that, like patterns, symbols and colors attract, resulting in invisible sympathies of harmony—a spiderweb of energies being drawn to the integrating object, resulting in more potency. Contagion is based on the belief that once something has come into contact with something else, a lasting harmonic linkage is established. Skin, bones, and feathers from predatory animals wrapped on the shaft call in the qualities of these creatures, increasing the success of the hunter and his weapon.

A weapon, created through a ritual process and designed with magical art, is a symbol of the unity of spirit and material worlds. It exists in material form, but the shape of the device originates from mythic consciousness, preserving a sympathetic harmony with that mysterious other world. It can be used in both worlds with authority. It is then a talisman and an amulet as well as a tool.

The value of a spear depends on the distance with which it can be used effectively as well as the sharpness of its point. In many cultures, people invented devices called atlatls, or spear throwers, to help increase leverage and throw spears more accurately. They first appeared in Eurasia thirty thousand years ago and then in North America about twelve thousand years ago. When Columbus landed he was met by Carib Indians who used atlatls.

The atlatl is a grooved stick about the length of the forearm that connects to the end of the spear. It dramatically increases the leverage that the spear thrower can apply to casting the spear. To add distance, the atlatl spear is shorter and lighter than the hand-held projectile, and it is called a "dart." Today there is growing interest in reviving hunting with atlatls, as with practice a person can cast one fifty yards or more with considerable accuracy and potency. A five-ounce dart can be cast at one hundred miles per hour, delivering it with as much potency as an arrow from a hunting bow at short distances.

One of the great synchronous events in the evolution of human consciousness was the invention of the bow and arrow. Before modern communications technology, people all around the world independently discovered that by bending one stick or spear and connecting the ends together with a cord of animal or vegetable origin, a spear could be projected considerable distance. Paintings and carvings in Paleolithic caves in Spain dated twenty-five thousand to thirty thousand years ago show people hunting with bows and arrows. Some experts speculate that Neanderthal people one hundred thousand years ago may have used crude archery equipment, but we have little evidence to support this theory. The best substantiated evidence of archery in use is in 13,000 B.C.

The oldest bows tended to be made of wood or bone and sinew, all of which may be eaten by rats and mice and decomposed by the forces of nature, which make dating difficult. The only major group who did not develop archery for war or hunting appears to be the Australian aborigines; instead they invented the boomerang and used atlatls with great accuracy.

The bow and arrow was the principal weapon of war in modern times for more than three hundred years. In sixteenth-century England, it was mandatory for every able-bodied man to own a bow and arrow. Some of the greatest early archers were the Turks, Egyptians, Vikings, Greeks, and Chinese. American Indian bows and arrows tended to be smaller and less powerful than those in other parts of the world, but Indians' skills in tracking and stalking made them equally effective hunters.

For many cultures, the materials from which the bows and arrows were made had a spiritual connection to the spirit of the bow. Making a bow from the antlers, horns, and ribs of deer, elk, reindeer, and caribou, for example, was felt to increase the chances of hunting those animals with success. If the bow was made from wood, the species of tree was felt to influence the spirit of the bow. In Robert Graves's *The White Goddess*, a classic study of Celtic and Druidic lore in poetic form, the lines of the famous poem "The Battle of the Trees" describe nine magical trees of Ireland—oak, ash, willow, yew, whitethorn, blackthorn, sloe, elder, and the reed.[3] The reed was considered a tree, for it could be used to make arrows, and the yew was seen as a gift of the gods for its wood made the finest longbows.

American Indians made bows from osage orange, yew, and hickory, as well as the bones, horns, and antlers of deer and elk. The best arrows were made from cedar and pine, both of which are lightweight, easily worked, durable, and do not readily warp. My Lummi friend Kenny Cooper tells me that his tribe preferred yew wood for making bows. They began by finding a yew tree. Then they asked the tree's permission to use a branch for making a bow. This involved prayers, a ritual, and some kind of sacrifice, such as some food placed near the tree for the spirits. If the tree approved, then the process began by selecting an appropriate limb and placing weight on it so that over a year or two the limb would naturally curve into the desired shape of the bow. When it came time to cut the limb, the basic shape was already in place with the aid of the tree, whose spirit was then working with the archer.

Cooper also tells me that when he was taught to make arrows, there was a rule that if you shot an arrow at a deer or a bear, you could not retrieve the arrow if you missed. Aside from the miss being an omen that the spirit of the arrow wasn't working correctly, Cooper says it made the Lummi outstanding archers because making a good arrow takes a long time and a lot of hard work.

Archery is a frequent theme in the myths and legends of many cultures. The Greek god Apollo was their god of archery. Eros or Cupid shot a magical arrow causing him to fall in love with Daphne, the daughter of Peneum, the river god. Eros, it seems, did not want to get into a permanent relationship, and so he shot a lead arrow of hate into Daphne, causing her to hate Apollo and turn herself into a laurel tree. The root of the modern word *arrow* is thought to be Eros. It seems no mere coincidence that a hunter aims for the heart to execute a humane kill and Cupid also shoots for that same organ.

The Greeks also named a constellation in the southern skies of the Milky Way Sagittarius, which means "the archer." Typically, mythic archers are passionate people, using their arrows to effect change in the world and gain recognition for their prowess, such as Robin Hood. William Tell, the Swiss archer who shot the apple off another's head, apparently used a crossbow, not a longbow. Feats of marksmanship were once seen as a divining. In many lands, shooting arrows at a target, instead of at one's enemy, was one way of settling a dispute. The reasoning here is that someone who knows he is telling the truth will be dead sure of himself, whereas a liar will know in his heart that he is wrong and that awareness will cause his aim to waver.

Bows and arrows also are used by shamans for spiritual purposes. Mircea Eliade says that "the chain of arrows" myth is a worldwide common metaphor to describe shamanic soul journeys into the Upper World. Typically the shaman shoots "arrows," which are prayers, and if each one is true to its aim, it becomes a step in elevating one's consciousness until it reaches the desired level.[4] Among the Huichol of Mexico, prayers are symbolically depicted as arrows, and in the process of conducting a ritual, participants make "prayer arrows," which are sticks decorated with feathers, shells, and other objects and wound with colorful yarn. In a ceremony one might make a prayer arrow, weaving together thoughts of a prayer, and then place it next to the fire at night, hoping that the spirit of Grandfather Fire will carry the prayer on to the next world and result in a dream for the arrow maker. In other tribes, shamans might shoot arrows into the sky as part of a ceremony to make it rain or snow. Some shamans also speak of negative "thought arrows" that are mentally shot or thrown at people by sorcerers and cause disease when they enter the body of a person and disrupt his or her energy flow.

Flutes, drums, and rattles are some of the most common shamanic instruments; the rhythms they make are used to entrain brain waves and lift the mind into other dimensions. Studies show that drumming rhythms of four to seven cycles per second are used by many tribes to induce trance.[5] While a drum may have many symbolic qualities, a bow also can be used as a rhythm instrument. It is not hard to imagine a hunter sitting beside a fire at night, feeling a little lonely and wishing to call on the spirits to help guide his hunting, who would use an arrow to tap out a rhythm on the taught bowstring as an aid to a chant or song, or another bowman wanting to celebrate the day's success and beating out a rhythm on his bow as he danced around the fire in ecstasy. Eliade said that "the shamanic drum originally was used to drive away evil spirits, a result which could also be obtained by the use of a bow," among the Tartar and Altaic shamans.[6]

One evolution of the bow is the string instrument, such as the guitar, lute, dulcimer, sitar, or the violin, where the modern bow may have once been an arrow drawn across the string. It is possible to change the pitch of a musical bow by applying pressure to the bow, which changes the tension on the string. Some tribes made a shorter "mouth bow," which was played like a Jew's harp, one end of the bow staff being placed against the cheek, using the mouth cavity as a resonator. In Brazil, the berimbau is a single-string musical bow with a gourd at one end. It can produce a remarkable range of sounds through a combination of changing tension on the string, striking the string in different places with a stick like an arrow, and cupping the gourd resonator against the body.

Combining the bow with a drum results in a guitar or a lute. The steps to heaven become frets on a fingerboard, and the notes become arrows. Listening to the wail of a frenzied electric lead guitar, one knows that Eros is still watching over the use of the craft, perhaps from the constellation Sagittarius.

In the last few decades archery technology has undergone a tremendous transformation. The good old-fashioned Port Orford cedar arrows, which were the norm until the 1950s, have now been replaced by hi-tech shafts made from aluminum alloys, graphite, and fiberglass. You can still buy longbows and the more recent wood and fiberglass laminated recurves, but the first choice of most archers today is a metal compound bow. It is not a thing of beauty like its ancestors, but, using cables and pulleys, the compound bow enables the archer to easily draw bows of great strength with relatively little effort. This makes the compound bow an extremely accurate and lethal weapon, without the archer needing to be a body builder.

Arrows to shoot small game such as rabbits and quail use blunt tips, the killing power coming from the shock of the arrow striking. For larger game, however, the arrow must be tipped with a razor-sharp point with two or more cutting edges, turning the arrow into a projected knife, with the kill coming as a result of hemorrhaging, unless a nerve center is hit directly.

I still have the nearly forty-year-old fiberglass bow my father game me to hunt deer in Michigan. Today, when I hunt wild boar in the hills of California, however, I pack the state-of-the-art compound bow made from metal alloys. Up to forty yards, this contraption of pulleys, cables, and bent steel has the accuracy of a pistol, and I want to be sure that when I shoot, it will be a killing shot. In in the overall experience of hunting, proficiency in one's skill is as important as the kill.

Some antihunters decry bow hunting, claiming it causes too much suffering. It's true that sometimes you do not hit the animal where you want to, and then you have to try to follow a faint blood trail for half a mile or more. A redeeming feature of the bow over a gun is that the knifelike wounds of a superficial shot will heal quickly, whereas a bullet hole through a deer is much more likely to become infected and lead to a long, slow death if the animal is not hit in a vital area.

To answer accusations leveled by antihunters that the bow and arrow are inadequate to kill big game, the California Department of Fish and Game has undertaken an extensive review of the research available to see if this charge is substantiated. They find that the data do not support the accusation. In one research study, seventeen deer were shot in the chest cavity by broadhead arrows shot from a 60-pound compound bow. The average deer was immobilized within 30 seconds after the arrow struck. As a comparison, twenty-eight deer were shot in the chest cavity with a 30.06 rifle and the average time to immobilization was 22.3 seconds. Other studies showed that deer shot in the chest cavity with compound bows ran an average of 91.44 yards before dropping, while those shot with a 30.06 rifle ran 70 yards before dropping.[7]

Some native archers use poison, like curare or aconite, on their arrow tips to increase the chances of a kill even if a superficial hit is made. Usually the natives who use poison do not have bows and arrows that otherwise would have much killing power. All they have to do is hit the animal, and the poison does the rest. I can see the point of this, but personally I would want to be pretty sure that if I cut myself with that same arrow, I would not be the victim. The accuracy of a modern compound bow is such that if you are a decent shot, you don't need poison-tipped arrows to make a quick, clean kill.

I prefer to hunt large animals with a bow and arrow rather than a rifle. If there is a need for meat, then I will use a rifle, but otherwise it is the twang of the bowstring that calls me. I marvel at how a large animal can be dropped in its tracks at four hundred yards, but to me that is shooting, not hunting. With a bow in hand I know that I must get within forty yards of what I will shoot, and I must be willing to wait until I can take a clear shot at the vital killing area. I am aware that this handicaps me considerably over the hunter with a rifle, but I accept the challenge with relish. Being an archer requires me to enter a state of consciousness where mind and body are one and all senses are in high gear. Every step I take must be silent. Every sound I hear may be a sign. Even odors on the wind may be telltale signs of game. When I see a big boar or a buck, I must somehow get the bow drawn and the arrow shot without spooking it, enjoying the rush of adrenaline that comes while knowing that I must be still in mind and body in order to hit the mark in the kill zone. There is only the now, at the peak of the hunt, a natural high that requires me to set aside my ego and step into line with a heritage of hunters and warriors that traces back 5 million years to Olduvai Gorge. *Homo sapiens* is a pleasure-seeking animal, Freud taught us. We are drawn to sex with the payoff of an orgasm, and food with the pleasure of taste, as well as meeting the basic needs for procreation and nourishment. In the hunt there is another kind of psychological carrot offered, the hunter's high, which has an appeal above and beyond the call of the chore of putting meat on the table. It is this kind of built-in psychological pleasure that has guided the hunt for millennia, moving it from a matter of necessity to one of passion. It is not necessary for a hunter to kill something to have a satisfying day, but if there is not an occasional follow-through, the experience is diminished in intensity as well as meaning, as long as meat is a part of your diet.

Pebbles to Bullets

The first weapon was probably a rock, held in the hand and used to strike, killing by shock more than by cutting. In time, people learned to throw the rock with accuracy, increasing its striking range. One of the first technologies to increase distance was the development of the sling, which is a leather thong with a pouch that can hold a stone. The hunter swings the sling over his head and then at precisely the right moment lets go of one of the ends of the thong, releasing the stone. Vernon Masayesva, chairman of the Hopi in Arizona, tells me that when he was a boy that was how he hunted jackrabbits. I'm glad I don't have to hunt with a sling. Unskilled with such a sling, the hunter is in more danger than the hunted! A modern high-tech slingshot with tubular rubber bands that shoots ball bearings, on the other hand, is a very accurate and deadly weapon.

In time the thrown pebble was transformed in two directions. One of the first hunting technologies was a deadfall, a type of trap in which a large stone would fall on an animal who set off a trigger of some kind. With the aid of a lever, such as a catapult, technologies were gradually developed to cast large stones over distances. The other adaptation was to move to smaller pellets, which could be cast with more speed. While many bows were used to propel arrows, others were used to shoot round hard rocks. In southern Asia and South America, pellet bows were widely used through the nineteenth century. The bows look like conventional bows except that a pouch that can hold pellets is mounted on the string. The hunter casts a stone or other hard object by curving the horns of the bow outward and pulling the string slightly off to the side.

The pellet bow is very well suited for bird hunting, especially birds with unique feathers used for costuming or as ritual objects. Arrows kill by piercing the bird's skin and cutting, resulting in blood spoiling the plumage. Pellets shot from a bow kill by shocking, thus the feathers are not harmed. Until recent times pellet bows were used by the royal palace guards in Bangkok, Thailand, for crowd control without excessive injury.

Gunpowder is among the most important inventions in human history. A mixture of 10 percent sulfur, 15 percent charcoal, and 75 percent saltpeter (potassium nitrate), black powder is thought to have been developed in China around A.D. 1000. Its early uses were for bombs and grenades, as well as fireworks, which have a number of spiritual uses in Chinese culture. In Chinese fireworks factories, to honor the "spirit of black powder," workers are required to work in silence, paying tribute to the awesome power arising from this chemical invention. Firecrackers are used to drive away evil spirits and malevolent ghosts, such as during the Chinese New Year's celebrations. To protect homes and offices throughout the year, many Chinese hang decorative firecrackers over the entrance doorway.

By the thirteenth century, the Arabs had gunpowder, and Roger Bacon wrote about it in 1242. It was not long before people discovered how controlled gunpowder explosions in a hollow chamber could be used to propel pellets, thus giving birth to guns. After gunpowder, the next most important breakthrough in gun technology was the discovery that lead could easily be cast into bullets. By the early fourteenth century, black powder guns were widely manufactured in Europe. The introduction of guns into Europe was a primary factor in the destruction of the feudal system.

Gunpowder allowed pellets to be propelled with much greater force, accuracy, and speed and over longer distances than arrows. When its blast could be channeled into the barrel of the gun, it made the projectile fly in a consistent direction. Early guns were cumbersome, slow to load, and useless in wet weather. They were loaded through the muzzle, each component had to be packed down into the firing chamber by a long rod, could fire only one shot at a time, and required some kind of external ignition explosion to set off the internal explosive charge. Misfires were not uncommon, and for a time the bow and arrow remained a more reliable weapon, as armsmakers experimented with flintlocks and other difficult firing mechanisms. The great breakthrough in ammunition technology was the prepackaged shell, which internalized the explosion, drastically cut the time to load, and led to the nineteenth-century development of guns that could shoot a number of times in rapid sequence. The machine gun, which has probably killed more people than any other single weapon in history, was invented in 1885 by Hiram Maxim, an American.

Black powder created large clouds of white smoke, which quickly identified the shooter, left bore-restricting residue, and absorbed moisture easily. It was replaced by smokeless, nitrocellulose based powder early in the twentieth century. Today, black powder is used mainly for fireworks and for muzzleloading rifles and shotguns, which are making a big comeback as more hunters choose to hunt for the richness of the experience over the certainty of killing game.

There are three main types of guns that use shells; long-barreled rifles, which shoot a single bullet; long-barreled shotguns, which shoot many small pellets over shorter distances; and handguns. The value of the longer barrel and the shoulder-rest stock is greater accuracy. Modern hunting rifles equipped with telescopic scope sights are deadly accurate up to four hundred yards. A shotgun loses all of its killing power before the cloud of tiny pellets has traveled the length of a football field. Handguns are not as accurate as rifles, but they are more maneuverable and can readily be concealed. Some hunters hunt with handguns, but the vast majority of handgun owners use their weapons for self-defense or recreational target shooting, unless they are criminals.

In January 1981, President Ronald Reagan visited Mexico and gave President José Lopez Portillo a present of a Remington rifle, to "set the tone for a friendly relationship." This gift created quite a stir, especially among gun control advocates, but among hunters, and Reagan is a hunter, such a gift is a real honor. Hunters care deeply for their guns, sometimes engraving rich patterns and scenes on the metal barrel or in the wood of the stock. Today, such scenes are considered decorative, but in earlier times when magical thinking prevailed, the artwork on arms invoked spiritual support for the hunter.

Metalworking was once a spiritual craft. Smiths—ironsmiths, blacksmiths, gunsmiths, and the like—were once seen as wizards of alchemy who used magical spells, as well as considerable skill, to impart spirit into steel, making a cold, hard object take on soul. The Finnish mythic hero Ilmarinen's forge fashioned swords of extraordinary powers and a magical box called a *sampo* that was a cornucopia of good things. In West Africa, in the religion called Ifa of the Yoruba nation, the god, or *orisha*, Ogun presides over hunting, and Ogun is the Spirit of Iron. While hunting is important for food and clothing, initiates in the Order of Ogun also have a number of other community duties including metalworking, circumcision, barbering, conducting ritual sacrifice, monitoring priests to be sure they are honest, defending the village, and enforcing capital punishment. According to Ogun initiate Awo Fa'lokun Fatunmbi, the common thread that binds all these diverse tasks together is the desire to know the truth.[8] Ogun is also a diviner's god, and protects the pearls of truth once discovered with the coldness of the steel blade of his symbol, the machete. Those who seek the truth, thus serving a higher purpose, are said to follow *iwa pele*, which means the road to "good character." Following this path results in wisdom, for in traditional Yoruba culture, *iwa pele* arises from knowing *awo*, which means understanding the invisible forces that sustain and form nature. *Awo*, Ogun's initiates assert, cannot be known by the intellect alone, which is similar to the Chinese concept of following the Tao, which also cannot be understood by intellectual means.

Metal weapons require precision. The better the parts fit together, the more accurate and potent the weapon becomes. A truly fine handcrafted gun has the qualities of a clock that never loses time, a razor-sharp knife that cuts through bone, and an arrow from Cupid's bow that strikes the exact center of the heart. Making a truly outstanding gun is among the most skilled crafts in the world. Its balance and feel in the hand command respect in the same fashion as a precision sports car does when you touch the wheel. When one creates with such precision craftsmanship, the resulting product is art.

In modern society, art has waned as our lives have become more mechanized; and this is a source of much psychosomatic disease. The cure for alienation is to discover one's feminine side— emotions, feelings, and intuitions. Common approaches include therapy, consciousness-raising groups, drugs, and meditation. Men (and women who have fallen into conventional men's roles) who have undertaken this journey get in touch with buried feelings but often report feeling a combination of vulnerability and rage. The problem here, psychologically, is that stripped of their defensive emotional armor, they find that they have no internal guidance system, and so to defend themselves they become hypersensitive and defensive, and/or self-righteous, even pompous. Personal growth is a long, slow process, with many unexpected turns. At the core of the masculine principle is the cold knife edge of truth. Among traditional cultures, men cannot use certain dangerous weapons until they have undergone full apprenticeship and learned to discern the right times to use these weapons. The rationale is straightforward: If you want to use a lethal weapon, you first must learn when to use it, as well as master the skills of its applications.

The rapid urbanization of society, the decrease in leisure time, the pressures of work that increasingly pull people away from physical labor, and the loss of natural areas have tended to erode the spiritual legacy of the hunt. Ogun is not just a West African deity. His order represents a mythic theme whose spirit activates an essential archetype of human nature that is endangered in modern society—knowing truth. We are wonders at obtaining information, but the skill of sifting through the flood of words and ideas to get down to what is valid is not cultivated. For the sake of saving time, many choose to become politically correct. If we could grasp the spirit of iron, know how it permeates a weapon and makes it become possessed with spirit, misuse of weapons of all kinds would decline sharply; and warmth in human relations would prosper. In a similar fashion, hunting clubs and hunters' organizations, which are the counterparts of the West African Order of Ogun and the hunting societies of nearly all hunting cultures, need to accept more responsibility for keeping alive the spirit of hunting, which includes duties of conservation of wildlife, education of neophytes, protection of moral ethics, and sharing the success of the hunt with the community in the service of charity. It is a sad statement about the media's desire for sensationalism that more people are not aware that by far the largest environmental organization in the world is the National Wildlife Federation, which is a consortium of wildlife and sportsmen's organizations whose work is in keeping with the mythic themes of responsibility and service of earlier hunting orders. Instead we hear of a whole new array of groups claiming to be environmentalists, and in the process the respect that once was attached to the term has dimmed considerably.

A gun is a numinous object. Yes, it may have sexual symbolism, but beyond its erotic character, a gun has other meanings. When an honored soldier or statesman dies, guns are fired as a salute in honor of the deceased. Once it was felt that this would aid the soul of the person on the journey to the next world. In earlier times, guns were fired to encourage rain to fall, making their report a message to the gods. If a gun saves your life, or kills an elk or a bear, it becomes a valued friend, your closest companion in some situations. A gun is also a symbol of authority and law and order, or criminal extortion, murder, and mayhem. A gun is an archetype of potency, and archetypes themselves are neutral. How they are used determines whether they are good or evil.

When you shoot a bow and arrow, you must draw the arrow, hold it and then release. There are sights that you can use, but still many factors are left to judgment. With a gun, you load the weapon, take off the safety, aim, and squeeze the trigger. Guns are colder, more mechanical, and go off more easily than the bow and arrow. With any weapon, accuracy is based on concentration and skill, but guns can be fired by anyone much more easily than a bow. Some guns can go off when they are dropped. A bow cannot. A bullet may carry a mile or more. An arrow falls to earth within sight. Guns, some say, are an equalizer. Not hunters.

A gun in the presence of other people is a focal point of attention. Safety is always a concern, but guns represent the powers of life and death, and as such they demand respect. When we lose our respect for guns, we get into trouble, for it means we have lost respect for life.

Weapons, Violence, and Love

Hunters who hunt with guns of any type today find their pastime challenged on two fronts: animal rights antihunters and gun control advocates. As I have been working on this book a number of people have asked me how I respond to the antigun forces who feel that guns are a primary cause of the violence so epidemic in the world today. As I pored over research studies, and arguments pro and con about guns, I recalled an afternoon I spent in Detroit in the late 1960s. I was part of a team from the University of Michigan's Institute for Social Research, told to gather public opinion about the causes and prevention of violence for the National Advisory Commission on Civil Disorders. Our assignment was to interview leaders of block clubs, and the area that we drew was not far from where the 1967 riots had ravaged the Motor City.

I knocked on a door and was soon greeted by a chunky, white-haired man, perhaps fifty-five years old. As soon as I entered the room, I knew this was going to be a special interview. The walls were covered with souvenirs that he had collected during World War II. A number of rifles, swords, and articles of military clothing were displayed, along with pictures of his army squad. His home was like a war museum. We sat down, and I took out my interview schedule. The first question was "Do you think it is ever right to commit a violent act?" Asking this question seemed ridiculous in this situation, but for the sake of research design, I asked it.

"No," the man answered with a straight face. I almost fell off my chair. Here was a highly decorated soldier who had fought a number of battles and presumably killed men in combat, and yet he felt that he had never committed a violent act.

I put aside the survey papers and learned over the next hour or so that "violence," in his view, was an illegal act, a crime. Self-defense, or defense of one's country, was not illegal, and so war was self-defense, not violence. The key to determining what is violence, he told me, is the motivation of the people involved, and the laws that apply to the time.

Murder statistics compiled by the FBI and police departments show that the majority of the thirty to forty thousand murders every year in the United States are committed by relatives or acquaintances. Looking at murders over a forty-year period, the FBI reports that nearly three-quarters of them nationwide are committed by people who have had prior arrests for violent felony and burglary. A review of police records from Detroit and Kansas City given at the American Society of Criminology in 1987 reports:

> In ninety percent of the cases of domestic homicide, police had responded at least once to a disturbance call at the home during a two-year period prior to the fatal incident, and in over half (54 percent) of the cases, they had been called five or more times.[9]

More than 60 percent of the homicides committed in the United States in the last thirty years were committed with firearms. The NRA tells us that "if guns are outlawed, only outlaws will have guns." The statistics seem to bear this out, for 99 percent of the people who own guns never use them to commit violent crimes (unless you are of the opinion that hunting is a violent crime). Instead, it is fairly clear that guns with which crimes are committed are used primarily by people who have a tendency to be violent in their relationships with other people. Many conservatives laughed at California's infamous Self-Esteem Task Force, but the intent is right on target for reducing violence. People who have violent tendencies are usually insecure. Feeling that they have little personal power, they seek to compensate by resorting to violence to intimidate people so that they can be in control.

I have not been able to find any statistics about correlations between crime and hunters; I think there would not be any positive statistical relationship and more likely you would find that hunters as a group are less likely than others to commit violent crimes, even though they own guns or others weapons. Here I return again to Erich Fromm's studies of aggression, where he concludes that the motives of the hunter are psychologically very different from those of criminals. Criminal aggression, in itself, is different from self-defense, for it strikes out against people with an intent to do harm for personal gain and with little or no regard for the welfare of the victim. Its motive is to gain control of people, goods, and situations, and often violence is the threat that makes control possible. Self-defense is protection of self and/or the community in the face of threat. Its purpose is to stop fear, while criminal aggression seeks to create fear.

As I said earlier, the extensive research of Fromm and other behavioral scientists clearly finds that hunters tend to be peaceful people. While hunting seeks to inflict mortal damage using some of the same motor actions and physical apparatus as conflict, the mood is pleasure-seeking, even spiritual, and concerned with skilled challenge, and the results are food and other physical remains that are used for art. Hunters may be initially guided by the desire for recreation and personal enjoyment, but the research clearly shows that in time hunters tend to develop a deep personal fondness, even love, for animals and nature. Career criminals, on the other hand, tend to become hardened and show increasingly less compassion for their victims.

What we need to be concerned about today are the forces that move people to become violent, as much as getting guns out of the hands of criminals and reducing crime of all kinds. We are flooded with reports of violence through the media, which often slip into sensationalism to increase the audience. Each fall, when hunting season opens, the media quickly report hunting accidents and fatalities. The story they never tell is how few injuries and fatalities result from hunting in contrast to the carnage in an average day in every big city in America. Hunters tend to be stigmatized by the press, as well as by the antihunters, because the hunters' weapons are in plain sight. Imagine for a minute what the world would be like if all weapons in an urban environment were visible. I suspect that the media tend to stigmatize hunting because their job often is to sit in the middle of an information avalanche, and it is more convenient to be politically correct than to ponder the criticial issues of each news item. Sportswriters, especially those who focus on outdoor sports, need to try to expand their audience and help the public learn to think critically about hunting and what the real issues are. For the committed sportspeople, tips and stories are fine, but the people who do not hunt or fish need to know the critical questions to ask when they see a news story that challenges hunting. What is the credibility of the source? What are the real biological, social, and legal facts? We desperately need, more than ever before, to teach people critical thinking skills; the worldwide information network that serves up news threatens to mire us in a well-informed swamp.

On television and in movies, we are shown a world where violence is seemingly never-ending, and seldom do people use nonviolent methods to solve disputes. Heroes, we see, are violent people. They take the law into their own hands, because the cops, the government, and other people in authority are not competent. Movies and television can teach us how to handle violence realistically. Sometimes this means showing people how to resolve disputes without killing each other. On the other hand, a movie like Steven Seagal's *Under Siege*, in which terrorists take over a modern warship, shows us in blunt terms what may need to be done in situations involving terrorism. As counterterrorist expert Gayle Rivers stated very eloquently in *The War Against the Terrorists: How to*

Win It, sometimes the only way to stop terrorists is to kill them without any remorse whatsoever. We need to be taught how to think critically about violence of all kinds and what is needed to stop it. Perhaps the biggest mistake of modern entertainment is the glamorization of violence.

At the end of many movies, we see credits assuring us that no animals have been abused, which is important, but in contrast to the world of television and film in which I grew up, where the African safaris of big game hunters were prime time shows and everyone cheered for the hunters, today we almost never see animals killed on the screen. We need to ask ourselves why we feel it is okay to blast humans to pieces using methods that would not be condoned in real life. These images suggest to viewers of all persuasions that violence is an acceptable way to cope with problems. Yet we still reinforce the belief that animals never die by human hands—while we sit and eat fried chicken or steak. The movie *A River Runs Through It*, based on Norman Macleans's book of the same title, finally let people see the human side of fly fishing, kindling a dramatic new interest in the sport. We saw fish being caught, killed, and creeled. No punches were pulled. Some animal rights people may have been offended, but the vast majority of Americans loved the film, and it has won a number of awards. It became an overnight classic, giving insight into a sector of life that has been far too stigmatized. It helped us see how fishermen are human beings. I'm waiting for a similar movie to be made about hunting, showing that not all hunters are bad guys.

An important essay about violence in modern society that deserves careful study by concerned citizens is the book *The Culture of Complaint*, by veteran journalist Robert Hughes. Hughes compares our present society with that of ancient Rome, when people lost touch with the guiding myths and slipped into overintellectualism and hubris, setting the stage for the downfall of that civilization through violence. In a world where mechanical aggression and mass society prevail, sensationalism, extremism, and terrorism emerge as ways to gain acknowledgment and make people feel more powerful. A case in point is the animal welfare movement, in which there are millions of good, honest, law-abiding people who want to help animals, but the people who tend to get all the press are ones who thrive on fueling conflict, controversy, and hate. They do so by feeding people's feelings of powerlessness, stirring up repressed feelings of frustration and pain from all kinds of human situations beyond animals, and mining unconscious hatred to fuel and fund their antics. If a group's identity is based on fueling conflict among people, then it is adding to the already high fear and mistrust level in society that pushes people to become violent. Epidemic violence is the product of an emotional tone that emerges in mass society from the collective results of many forces, direct and indirect. All forces involved must be analyzed to develop systematic strategies (rather than Band-Aid attacks on symptoms) to curtail violence.

Reducing violence in hunting and animal welfare will occur when groups working in this area agree not to polarize and stigmatize each other but come together to find common goals in support of animal welfare. I know of advocates of hunting who are forced to carry sidearms because of threats against them and their families from animal rights advocates. I also know animal rights activists who carry guns, but not due to threats from hunters. There is food for thought in this imbalance.

There is considerable evidence to support the thesis that aggression of various kinds is an elemental part of human nature. Adrenaline, testosterone, and other chemicals that stimulate aggressiveness flow throughout our bodies. War, play, dance, art, exploration, making love, and even the human need for exercise to stay healthy are dependent on action. Without aggression, we probably would have become extinct long ago. In modern society we have lost touch with our biological, chemical, and emotional nature and called the loss "normal." We have created a world where skyrocketing inflation is called "prosperity." Despite the millions of tedious, mechanical jobs involved in running the Information Age, we speak proudly of "employment" as if all jobs are equally valued. What is missing from our usual analysis is an assessment of the underlying emotions generated from each of these conditions. Humans can adapt, for short periods of time, to situations that are not pleasing to them or even in their best interests. Forced to conform to these conditions over time, eventually the pot boils over into outrage. The physical fitness movement that has mushroomed in the last twenty years is one of the best signs of hope for reducing social tension as well as keeping fit. Then, just as we start to realize this, schools start taking away physical education classes that can teach lifelong recreational skills as well as competitive sports.

When it is easy for guns to fall into the hands of frustrated and frightened people who feel they have little power, violence escalates. To reduce violence, we have to make sure that people who use guns have the skills and respect necessary to use guns safely, legally, and sanely. To buy a gun to replace my treasured old 12-gauge double barrel, all I had to do was have enough money and the proper identification and be willing to wait two weeks while the state checked to see if I had a criminal record or a history of mental problems. (I passed, incidentally.) California has had stricter gun control laws than those imposed by the Brady Bill for years, and their effect on reducing violence has not been great. The 1992 California murder rate again set a record—3,920 people were killed in homicides. It is interesting to note, however, that the ratio of 12.5 murders per 100,000 population is essentially the same as the previous year. Slightly less than 73 percent of the murders were committed with guns, and about half were committed in Los Angeles County. Of those killed, 82.1 percent were males, 43.2 percent were Hispanic, 27.5 percent were black, and 23.4 percent were white.

To get a California hunting license, people have to pass a hunter safety education course that involves a minimum of ten hours of class time. Since this was made a law in 1954, hunting accidents of all kinds have declined substantially. I would like to see gun owners have to pass similar classes to show they know how to use weapons before they can buy one. I would not be opposed to mandatory gun registration. It would seem to separate quickly the illegal guns from the legal ones.

Assault weapons are not the tools of a hunter. Anyone who needs rapidfire weapons that spit out several bullets per second to shoot a deer or an elk is not hunting. Uzis, AK-47s, and other automatic weapons are designed for combat—to intimidate people as well as to kill—and they tend not to have accurate aiming devices. Someone learning to hunt with an assault rifle or handgun would likely count on multiple shots to hit something rather than take the time to make a single, clean, killing shot in the vital zone. These weapons would promote taking "sound shots" and then seeing what you hit. My personal rule is that if you don't think you can kill an animal cleanly with one shot, don't shoot.

We cannot simply put the lid on human nature and declare that we do not have aggressive instincts. Some of the people who declare that they are peaceful and are working for peace sometimes seem to be among the most aggressive. This seems especially true of some of this decade's politically correct types. Deny the shadow and in time we become the shadow. We need to find ways to express our aggressive energies in a creative fashion along the lines of Aldo Leopold's land ethic. Hunting happens to be one of those natural ways in which humans express aggression in a manner that benefits themselves and the species. Not only does it provide good recreation and healthy food, but the act of hunting itself can be psychically healing for people. As a young boy, Theodore Roosevelt had asthma. It was not curable by conventional medicine, so the family finally sent Teddy to a European nature spa for a cure. After several months his asthma went away, and Roosevelt then went on his first African hunting expedition. He came home cured. For the rest of his life, whenever troubles clouded his life, Roosevelt's cure was hunting and fishing. He came to know the importance of catharsis as a cure, and forever after he was grateful to nature for his health and showed his thanks by founding the modern conservation movement.

When people come together to form community they create agreements about how they will relate to one another. Laws are necessary to protect individuals, but law enforcement is most effective when it arises from respect and negotiation, not threat of punishment alone. One way in which all groups concerned with animals and nature, hunters and nonhunters alike, could make a step toward reducing violence would be to join together to support efforts to curb poaching. It has been shown in many situations that as soon as legal hunting declines, poaching increases. In California now there are more deer killed illegally than legally. You will never stop people from killing animals by outlawing hunting. But you can stop the unregulated killing of animals by calling for all people who care about animals to join together and put poachers out of business. For many species, that would prevent more animal suffering and cruelty than stopping legal hunting.

Some states have toll-free numbers to call to report poachers. In addition, there is a national number, 1-800-800-WARDEN. I feel it helps to have a visible hunting license with numbers on it required for all hunters, so you can report the identity of lawbreakers without needing to get next to people. Some people could abuse this, find out your number and harass you by calling in false claims, some hunters respond. I agree. This is just one more example how animals suffer from hunter harassment. Phones that take incoming complaint calls, however, increasingly have devices that will instantly tell them the phone number of the person making the call.

What is especially alarming about guns and violence today is the number of kids who use guns for purposes other than hunting. I had guns when I was in my teens, used them almost every day in October and November, and so did a number of my friends. We had guns to hunt with. Today, when fewer kids hunt, they own guns and increasingly point them at one another. I wonder to what extent we could reduce violence by getting more kids out hunting and fishing. Putting kids and guns together does not automatically result in crime. The social conditions are what move kids to crime. Former President Jimmy Carter, an avid outdoorsman as well as a peacemaker and deeply religious man, describes his weapons in his youth:

> By the time I was ten years old I had graduated from flips, slingshots, and BB guns to owning a .22 rifle and a bolt-action 4-shot model .410 shotgun from Sears Roebuck.[10]

In November 1993, Florida passed a law prohibiting juveniles from possessing firearms except while hunting or target shooting under adult supervision. Penalties for violating this law are for both kids and their parents. Similar laws are now in place in Colorado and Utah. It makes a lot of sense, especially in a world where people live much closer together. When I was growing up, for quite a while the closest other house was a quarter-mile away up the only road in the area, and we lived at the end of the road. If you wanted to go any farther in the direction of the road, you needed a boat.

A gang is a group of kids who care about one another. So is a team. Gangs form for protection when kids feel abandoned. You can't take away the need for kids to have friends and get together in groups, but you can give them more opportunities to come together and accomplish goals that are valued by the community. Drug use has become so common because people do not know any other way to get high, and because life for many kids is not a pleasant experience. Soaring rates of single-parent families, in which kids are often left to fend for themselves far too early in life, are a contributing factor in violence in our streets. Homicides from gang violence in 1992 accounted for 22.2 percent of the murders for that year in California, up 400 percent from the 1983 toll of gang-related murders.[11]

The guns used to commit gang violence are not generally ones purchased at legitimate gun shops. They come from thefts as well as the thriving black market for arms and ammunition, which ought to be a principal target in violence control. Schools will no doubt need to install metal detectors in many more cases, and police will need to have devices that can sniff out guns at a distance; but we will gain real control of violence only if we seek to prevent its causes, which lie in the social conditions that breed crime.

We need to look at why kids feel moved to form violent gangs in the first place, and our schools are a good place to begin. Schools have all too often slipped into "worksheet wastelands," to use the words of psychologist Thomas Armstrong, where getting high test scores has become the holy grail of education. Faced with budget cuts, many schools have abandoned classes that build self-esteem and promote health, including art and physical education. Science and math have slipped into the domain of advanced math, which few people ever see, let alone use, after graduation. Homework has become a way to keep kids under control, establishing a pattern in later life that says only workaholics succeed. Is it any wonder then that some of the most frequently prescribed drugs in the United States are for hypertension and ulcers?

The rampant cynicism that is the emotional tone of many kids today tells us that fear of the future is the real emotional issue for them. They attend a school system that has gotten out of the control of their parents and the community. It is run instead by college entrance administrators. And then look at the world the kids are facing—crime, overpopulation, terrorism, environmental crises, AIDS, inflation, and more. Nice future to look forward to. Schools do precious little to build a sense of vision and hope for kids. Violence grows when hope wanes. Education ought to be based on developing what is needed to become successful in life, cultivating skills that are practical, as much as accumulating knowledge that is a prerequisite for college.

One simple way to help reduce violence is to look at how we can improve the criminal justice system. For dangerous felons, I think "three strikes" is too long to wait before being called out. On the other hand, there are people who fall into crime who can be rehabilitated. I am not convinced that incarceration alone is particularly useful in preventing crime once the convicts are released back into society. For first offenders of minor crimes, proposed rehabilitation programs that have a "boot camp" kind of physical conditioning program coupled with developing mental discipline seem to offer a positive alternative to simply locking up a person for punishment. Addicts of any kind usually have a problem of low self-image, and heroin addicts tend to have very low self-esteem. Working with heroin addicts, my good friend Thomas Pinkson developed an outdoor challenge program, using wilderness skills like mountain climbing, white water canoeing, and wilderness solos, which resulted in significantly lower recidivism rates than any other treatment program in the San Francisco area. Nature can be a great teacher and healer, if we give people the chance to learn how to appreciate nature's powers. Pinkson notes that for many of the hard-core drug users in the program, being out alone in a natural environment was more frightening than walking inner-city streets. Years later, some of them periodically call him to thank him for helping them get off the vicious cycle of addiction.[12]

Symbolically, weapons are masculine. Their potency comes from being cold, hard, and precise. The excess of the masculine mode of consciousness is to lose touch with feeling, becoming cold and hard and unemotional so that values no longer mean anything. The extreme of this would be the psychopathic killer who feels no remorse whatsoever for killing anyone or anything. The myth of Cupid shooting an arrow into a heart to inspire love, however, shows us the natural wisdom of the human psyche, for as one pole is reached, and truth is spoken, hitting the bullseye so to speak evokes the balance. Telling the truth is the cornerstone of personal intimacy, which is the root of trust and love. Where there is balance in life, there is health. Man is a tool maker and a tool user. As society evolves, we also have to increase taking responsibility for using the tools we make. With weapons in hand, let's turn to the feminine side of hunting for balance.

If one really wishes to be a master of an art,
technical knowledge is not enough. One has to
transcend technique so that art becomes an
"artless art" growing out of the unconscious.

In the case of archery, the hitter and the
hit no longer are two opposing objects, but
are one reality.

D. T. SUZUKI[13]

Seven

The Feminine
Side of Hunting

Nature is harsh, severe and cruelly revengeful.
There is neither judgment nor rule, but the
revenge of the dark aspect of the feminine.
MARIE-LOUISE VON FRANZ[1]

Nothing happens by chance. In the middle of writing this book,
I found myself in the tiny village of Danville, Vermont, where
I was to be the keynote speaker at the annual convention of the
American Society of Dowsers. This is perhaps the only environmental organization whose purpose is to help people cultivate a
more sensitive relationship with nature. I am not an expert dowser,
but for the last twenty years I have been studying the psychology
of how and why people love nature. The dowsers invited me to
speak because my research shows clearly that people with the
deepest nature kinship generally don't come to love nature just by
reading about it, or even taking classes in ecology. The root of the
love of nature more often lies in positive emotional experiences
that one has with the natural world. Magazines, books, and the
other media inform us, entertain us, and may help persuade us to
spend more time outdoors, but those who have a serious commitment to conservation generally begin to develop their ecological
conscience as children, when they first explore the fall woods as
the red and gold leaves pile up, traverse a frosty hillside covered
with powder snow, walk through a spring meadow filled with wild-

flowers, and pick blackberries beside the creek where the bullfrogs croak on a summer afternoon. Early hunting and fishing outings, for many people, are among their most treasured outdoor experiences, establishing a foundation for lifelong interest in nature.[2]

Dowsing is the ability to locate objects and conditions at a distance without the aid of our ordinary five senses. The archetypal image of a dowser is a crusty old farmer with a forked willow wand in hand who walks the countryside until his stick suddenly bends down to indicate that there is water underground. Divining, water witching, and other forms of ESP are challenged by the modern scientific paradigm. A hard-minded colleague of mine once walked into one of my classes as someone was trying to dowse and said, "That can't happen." Funny thing—before she came in and said that, nearly every person in the class had walked across the wooden floor blindfolded, dowsing wand in hand, and located the water pipe under the floor. But as soon as the statistician declared that dowsing could not occur, no one else in the class was able to get a positive result.

There has been enough research on dowsing and other forms of ESP to show that often the attitude of the experimenter is as important to the test results as is the prowess of the experimenter. When open-minded scientists study dowsing, such as quite a bit of research sponsored by the U.S. Office of Water Resources Research, they find that most people do seem able to locate underground water, pipes, and so on with a body sense. Stanford University engineering professor William Tiller feels that the famous willow wand of dowsing is really a "biomechanical transducer," a device that amplifies physical sensations perceived by the unconscious mind. It is not the wand that dowses, but the body and the mind. The wand simply helps clarify unconscious perceptions. Uri Geller, known for his psychokinetic spoon-bending prowess, has quietly become a dowser in great demand worldwide by businesses, government officials, and even military leaders wanting him to use his hands to find water, oil, minerals, and many other fascinating substances hidden to the five senses. Dowsing is an anomaly for modern society, something that isn't supposed to exist. Nonetheless, it is practiced by thousands of people, often with great success.

Carl Jung believed there are four primary psychological functions: thinking, feeling, sensate-practical, and intuition. Dowsing is a form of intuition. "Women's intuition" is what people often call it. Intuition involves knowing what is about to happen or sensing what cannot be known by the accepted five senses. Intuitive flashes, leaps of perception of truth arrived at without analysis or use of the five senses, defy rationality. This does not make them wrong but rather shows a blind spot of the modern definition of the mind, as well as a shortcoming of the Newtonian-Cartesian model of reality. Such ESP humbles hard-minded scientists. The purpose of intuition is to allow us to make quantum leaps in awareness, and the purpose of rationality is to make sure that intuitives are honest.

Of Jung's four functions, thinking and sensate-practical are generally seen as "masculine," while feeling and intuition are considered "feminine." The terms are symbolic, not actually limited to each sex, for, as Jung pointed out, in each of us there is a part that is of the opposite sex. Men have an "anima," a feminine side with emotions, intuitions, and feelings. Women have an "animus," which governs rational thought and physical aggressiveness. Dowsing, as an intuitive perceptual skill, is problematic for the field of modern psychology, because the academic community generally defines psychology by research models that can be most easily studied by mechanical models, which are predictable and replicable. Intuition is neither; it makes unpredictable leaps of faith, sometimes hitting the mark directly while other times being far afield. Some assert that the conventional theory of psychology arising from rationally dominated research methods is biased toward the masculine aspect of consciousness and therefore is discriminatory toward women.

Emotions and intuitions can't be readily quantified because they are situational as well as personal. In psychology they are more the domain of the psychotherapist than the academic psychologist. People go to therapists because their conscious minds are out of harmony with their unconscious and dis-ease and anxiety are occurring as neuroses or more serious psychoses. In therapy, clients unearth repressed emotions to release tension and fear and seek to understand what they really need and want, reuniting ego and unconscious to clarify self. Therapists work more with the feminine mode of consciousness, while educators work more with the masculine, at least in the "civilized" world.

In cultures that are more intuitive, such as native cultures, psychological difficulties are not as often due to repressed emotions, for such people live in much closer contact with their unconscious. They do not have to control themselves to conform to cultural norms as much as most of us do. Native peoples have their psychoses and personal problems, of course, but often their mental dis-ease is caused more by confusion about loss of one's intuitions, which serve as the primary guidance system for life, instead of clocks, schedules, and calendars. In native cultures, the role of the therapist is that of a witch doctor who typically drives away the external evil forces that are somehow impeding one from following his path, and then uses ceremony and ritual to help the person restore his connection with his higher self—i.e., spiritual guidance and intuition. Among the Huichol of Mexico, for example, to reduce tension and anxiety, sometimes the shaman does not treat the patient directly at all, but blesses his or her shoes. The symbolism here is that the shoes take one along a path, and the anxiety and confusion that a person feels is due to walking the wrong path. Blessing the shoes chases away the negative forces that keep a person from walking the correct path, being who they really are. Be wary of any psychotherapist who denies the validity of intuition.

After speaking to the seventeen hundred people packed into the Danville high school gymnasium, I wanted to sample some of the feelings about hunting among the locals who live in the ruggedly beautiful Green Mountains. Danville is one of those tiny, picturesque New England villages; one stop light, a village green, a

couple of restaurants and antique stores, a church or two, a hard-
ware store, and a general store that is the hub of social life. At the
general store, I bought some maple syrup and asked about hunting. "Oh, you gotta go see John O'Leary," the clerk said, and
pointed me north on the only road going that way.

Half a mile later I turned in at a white New England farm
house that had a gun-shop sign outside. The sign on the front door
said OPEN, and I walked inside. This was the parlor of a private
home. I thought there must be some mistake, so I backed out the
door, only to be met by a large farm dog. He thinks I'm a burglar, I
thought to myself, and wondered what I was going to do if the dog
agreed. Just then a muscular, white-haired man in a red-and-black
checkered wool shirt appeared from around the corner. "Hi," he
said warmly, "Come on in."

John O'Leary led me through the parlor and up the stairs to
the second floor. I found myself in a gun shop that had more arms
than all of Marin County, California, where I live.

John gave me a tour of the store. He has a story for every gun,
as well as a lot of good old-fashioned woods sense. Like most seasoned hunters, he echoed the sentiment that things aren't what
they used to be. Eighty percent of the private land in Vermont is
owned by out-of-state people, and access is getting harder to come
by, John said wistfully. "With so many divorces and women raising
the kids, the number of new hunters isn't what it used to be," he
added, "and you know, the worst part of it is that hunting is really
a family sport."

As if on cue, we heard a knock on the door, and John called
down, "Come on in." Up the stairs walked a lady with her son. She
wanted to buy a rifle. I sat back and watched. In a few minutes she
had selected a deer rifle and left. As she drove off, John turned to
me and said, "You know, women are some of the best hunters."

"Why is that?" I asked.

"They don't fool around. Some of these men come up here from
the city to get away from it all, and they sit around and drink and

play cards, maybe go out for a couple hours a day to hunt. Women, they don't get all caught up in that stuff. They just go out and hunt very seriously. Lucky, too. Intuition, you know. Women around here are usually more successful than men in the deer season."

Hunting is a cruel, perhaps even sadistic blood sport, some animal rights activists challenge. Merritt Clifton takes this train of thought even farther into the realm of sexual symbolism, claiming:

> Whether or not hunters shot deer to demonstrate sexual potency or out of sexual frustration, in symbolic lieu of raping and killing women, there can be little doubt that as a social ritual, much hunting is all about killing the feminine in the hunter's own self.[3]

Clifton's charge raises three issues about hunting that one finds peppered throughout the animal rights, feminist, and vegetarian literature. Charge one is that the symbolism of hunting is violently, negatively, sexual, portraying the victimization of women as well as being brutal to animals. This then makes hunters amoral and sexist as well as cruel. The second is that hunting destroys the feminine aspect of the self, regardless of the sex of the hunter, because hunters must deny their emotions and intuitions to kill. The third is the implicit assumption that only men hunt. All three accusations concern the psychology of hunting and so they are open season to close scrutiny in our quest to understand just what moves the hunter to take up arms and go afield to kill. The charges need to be considered seriously, for they represent a considerable amount of negative energy directed at the hunting community. Clifton's accusations pertain to women as well as to the feminine mode of consciousness, and so in my writing now I maintain my awareness that I am a male hoping to shed some light on the feminine—definitely a difficult situation. It is not easy for a man to present a psychology of women, but it is perfectly appropriate for a man to talk about a psychology of the feminine mode of consciousness. All men, after all, have an anima. I enjoy discussing this topic, because according to my scores on a Jungian type test, which describes our human ways of perceiving according to Jung's four functions, I come out as an intuitive thinker.

The Myths and Symbols of Hunting

As the hunter's moon rises in the autumn sky, a special feeling of excitement wells up from the depths of the unconscious. Sights, sounds, odors, and events all kindle memories tracing back to the Paleolithic that stir us to action. Jimmy Carter admitted after leaving office that in the fall when he was president, when he heard geese flying over at night, he would sometimes go up on the roof of the White House to enjoy the sight and sounds of the wedges of geese crossing the face of the moon.

Myths are symbols ordered together into stories—depersonalized dreams. They speak to the unconscious directly, bypassing the ego's need for rationality, and move minds and bodies to action. While in most cultures there are more men hunters than women, often the mythic figure who governs the hunt is female. Artemis in Greek mythology, Diana in Rome, Pinga or Sedna among the Eskimo, and White Buffalo Calf Woman among the Lakota are all versions of the same mother-goddess archetype. A strong and powerful woman, she presides over life and death and often is also goddess of the moon and guardian of the animals. As an older women, a mother of the animals, she is also a powerful witch who can cast spells that either favor or punish the hunter according to the respect shown for the animals who give their lives so that humans may live. Hunting magic is feminine.[4]

In high school, I was a serious competitive archer and at one point considered trying for the Olympics. An inspiration in this pursuit was a very attractive blond woman who shot in competitions with me, Ann Marston. Ann was Miss Michigan, a Miss America finalist, and a women's national archery champion. She was five or six years older than I and we were just friends, but in retrospect she was the essence of Artemis. She was vibrantly beautiful, very intelligent, and a damn good shot. Some people wondered how as a beauty queen she could also be an archer and a hunter. Mythically the answer is very straightforward. The ally or companion of Artemis is the leopard; the attractive young woman is accompanied by a killer cat bodyguard, or she is wearing a leopard skin costume symbolizing that the leopard is within her.

Psychologically, the combination of beauty and lethal potency makes a great deal of sense. Attractive people can draw all kinds of attention, sometimes more than they would like, and from people they would prefer to avoid. The leopard symbolizes a healthy animus for a striking woman; it serves to protect her magnetic beauty and allow her to go out into the world and be proud of her vibrant qualities.

The potential pathology of the masculine principle is to become cold, hard, unfeeling and without a sense of higher purpose. The result is a patriarch who either dominates others or suffers from psychosomatic illness, or both. Men can resolve these problems by going inward to their suppressed emotions and intuitions to discover feelings that had been denied or channeled into aggression. They can then shed their tough emotional suit of armor and learn to cry and to enjoy the simple pleasures of life, no longer needing conquests to feel worthy.

The potential pathology of the feminine principle is to become swallowed up in uncontrolled emotionality and sentimentalism, becoming either a victim or a hostile, defensive person who will not let people get close. Women can resolve this problem by becoming assertive about what they really want, including learning self-defense.

Mythic male hunting figures tend to be strong hunter-warrior types. Ogun in West Africa is a bare-chested muscular warrior brandishing a machete, who prefers wearing a animal skin and residing in the deep woods, but when he does show up in public, his presence is commanding and respected. He is powerful not just because of his physical stature and sharp blade but because his mind is one with nature, which is how he can discern the truth, determine justice, and hand out punishment. Ogun is the archetype of a wizard. He develops great personal power through his love for both the feminine in nature and the hardness of iron shaped into razor-sharp weapons. His psychological challenge is to speak and act from the heart, so he serves a higher purpose and is not swallowed up by the petty jealousies and revenge that result in unnecessary war and conflict.

Mythic figures are symbols, psychological types in the unconscious, both personal and collective. They activate personal symbols and energies and then give guidance about how best to express these sentiments in everyday life. One of the great gifts of the late mythologist Joseph Campbell was to help us recognize that each of us is a myth wanting to be lived in our own unique fashion.

Sigmund Freud, through his studies of dream symbolism, helped us see that instinctual drives of the unconscious are conceptualized metaphorically. It is our conscious mind that forms words, but the words spring from deeper sources. A woman client who came to me for help with exhaustion and a feeling of emotional numbness reported that she dreamed only in words and numbers. She was a successful engineer. Only physical symptoms finally made her realize what price she was paying for acting out a male stereotype. Starting her journey inward to discover her denied emotions, she began to have dreams in pictures. At first they were only black and white, symbolizing how her life was an either/or colorless existence. Then, in time, her dreams took on color. Eventually, she went back to school and ultimately became a very good healer. She knew from personal experience what the journey to self was like, for she had obeyed the first rule of becoming a good healer: "Heal thyself first."

I am reluctant, as a man, to say to much more about the symbolism of women's psychology, as women will insist, rightfully, that only they can best talk about the female psyche. So, I will now focus on the challenge that hunting represents to the symbolism of rape. In the crime of rape, a person, almost always a male, forces a female or another male to submit to sexual intercourse using threats of violence. Rape is a sadistic act. We have already shown, clearly, that the act of hunting, undertaken in a legal and ethical manner by a healthy person, cannot be shown to be sadistic; the motivations and biochemistry of a rapist are different from those of a healthy, passionate hunter. For a hunter to act with motivations similar to those of a rapist, he would have to first capture the animal, then bind it and submit it to painful torture, from which he would derive pleasure. The actual killing of the animal would be anticlimatic, and so death would be delayed as long as possible. The mind of such a person would be moved by an undercurrent of fear and rage fueled by a deeper sense of personal impotence—"I can't have you in a normal way because I am no good, so I will get you and make you pay for it, my pleasure coming from revenge, not erotic love." It is the violent force of rapists that demonstrates their underlying feelings that, without force, they will not be loved or sexually satisfied.

There are occasional cases of sadistic hunting. Not long ago a man in Oregon was convicted of buying aged circus animals, putting them in cages, and then charging "hunters" to shoot them for trophies. Accounts of this case say that sometimes the animals never even got out of the cages. This is one way that hunting could be like rape. As long as the principles of fair chase are followed, and the hunter seeks to kill each animal as quickly as possible, hunting cannot be defined as sadistic.

There is strong sexual symbolism in hunting. Hunters use weapons, such as rifles, shotguns, and arrows, all of which are thrusting, penetrating objects that, broadly interpreted according to Freudian concepts, have phallic symbolism. Stalking involves intuition, feelings, spatial awareness, and sensory awareness, which are feminine, but figuring out the maps, driving the vehicles, and chopping the firewood are masculine. The hunter, armed with weapons (masculine), enters into nature (feminine). One could see the symbolism of hunting as intercourse; if the hunter does so with reverence for nature, then you could conclude that the symbolic nature of hunting is that of making love. A kill is symbolically similar to orgasm, a peak ecstatic moment. Hunting has considerable sexual symbolism, in part because it is a basic instinct, like sex, which is implanted in our minds and bodies as a survival mechanism.

This analysis could go on and on. Freud helped us see how almost anything *could* have sexual symbolism. He also pointed out that people can evaluate sexual symbolism as positive or negative based on their unconscious feelings projected onto the environment, as well as any other interpretation.

There is no basis to conclude that hunting, per se, represents sexual sadism. Hunting, like any physical act, is an expression of personal potency, but behavioral research shows that the emotions and energies of mentally healthy hunters are positive. You can find individual hunters who are immature, even sadistic, but the vast majority are not and should not be labeled as such.

The Feminine Aspect of Consciousness and Hunting

When I was teaching at the University of Oregon I was introduced to a woman in her late twenties who had a serious physical problem. Every joint on the left side of her body was sore and swollen. She had been raised in the woods and her family was quite poor. Once they had run out of money for shells, so they had to hunt for food by traditional methods. She showed me scars on her arms where she had killed a deer with her bare hands, dropping down out of a tree and stabbing the animal with a knife. This kind of rough physical activity had given her marks on her body, but there had been no problem with her joints. When she was fifteen, her parents died and she was forced to move in with a family in town. Her foster parents were very conservative and religious. In a year, her left-side symptoms began to appear.

She was very bright, went on to college, and ultimately got a master's degree. All the while, her joint problem became worse and worse. Faced with operations and seemingly endless cortisone shots, she knew she had to do something radical. Working with her, we developed an intensive, radical program.

Initially she devised a twice-weekly strategy that began with a chiropractor adjusting her joints. It was extremely painful. Then she would come to therapy, where she would vent years of repressed anger by attacking pillows. Her dreams at this time were of volcanoes and earthquakes.

As the bottled-up rage was vented, the pain and swelling of her joints subsided and she began to examine her lifestyle. She had spent fifteen years living in a world where, aside from some schooling, the intuitive domain ruled life. The periods of the moon, tides, migrations of fish and waterfowl, and ripeness of berries had governed what needed to be done. Each season had its emotional qualities as well. Her illness, it seemed, had started when she suddenly had to cut off all the sensations and feelings that had been the core of her reality. Her cure was to recover them.

She quit her job as a mathematician, dug up the yard of her home to make a large garden, and grew enough vegetables to feed her family. Digging in the garden and chopping firewood began to take the place of pounding on pillows. Her husband rescheduled his vacation to the fall, and they went off to the coast to hunt and fish. During this period she would sleep only a couple of hours a night, her husband told me. He said that it was if she was possessed by the spirit of the wild itself. She took up weaving during the winter and began to sell her handicrafts. I am happy to say that in a year's time she was out playing basketball.

Research has shown that the right side of the body and the left side of the brain are associated with the masculine aspect of self—rational, analytical, linear thinking, and communication. The right side of the brain and the left side of the body are dominated by the feminine aspects—emotions, intuition, body sensations, spatial relationships, and so on. Her illness was repression of the feminine side of herself. Her symptoms were so dramatic because she had lived for so long in a nonlinear world, and she was an extremely intuitive person. You could say that she is an archetype of the price we pay for denying the feminine in modern society.

The more a hunter relies on stalking and on weapons with a limited striking distance, the more important to success is the feminine mode of consciousness. Sights, sounds, odors, and intuitions that come must be screened and woven into an awareness of the whole of a situation to make the connection between the hunter and the hunted. The mind-set of the hunter in action is unitive; otherwise the chances of success are slim.

Emotions are mental and physiological responses based on our subconscious appraisal of situations as being beneficial or detrimental to us. What makes hunting so exciting is the emotional intensity.

There are several emotional challenges to the hunter. The first is to overcome fear, as big game animals can be dangerous. Fred Bear, the archdruid of archery hunting, used to say that if teenagers wanted a real thrill they should hunt a bear with a bow and arrow as it would "cleanse the soul." Icy waters, storms, freezing weather, swamp muck, and rugged terrain, all associated with habitats where hunters must go to find game, represent threats, too. Knowing how to handle each of these situations builds maturity quickly.

The second emotional challenge is to control your excitement and make an accurate shot. Misses and wounded animals often result from jumpy nerves. You cannot know what it feels like to be a hunter watching a wary buck approach your stand, or to sit still when a hundred Canada geese are circling your blind just out of range, unless you have done it. You get a refreshing taste of this with a camera or bincoluars in hand, but it is not the same as when you are holding a lethal weapon that you intend to use.

The third challenge is to resolve feelings of guilt associated with the kill. One approach can be to steel yourself. Another could be to slip into a childlike innocence and see the whole thing as a game, not taking responsibility for what occurs. If hunters do not try to avoid feeling the full range of emotions involved in hunting and killing wild animals, however, they ultimately come to feel deep reverence for the animals they shoot. This is how hunting becomes a spiritual practice. The word *religion* comes from the root *religio*, which means to bind together and give comfort. Spirituality arises from the experience of the spirit; it is awesome, beyond words, and leads to the understanding of what reverence for life really means.

Writer Margaret Knox describes the emotional intensity of killing her first deer:

> My rifle crept to my shoulder. My pounding heart drove the sight in a circle around the doe's heart. Aeons passed before I steadied the barrel and fired. She flinched, ran a few steps, and collapsed. As I approached to deliver the last shot, her brown eyes glazed with mute reproach. I barely felt the kick or heard the rifle this time, but I turned cold suddenly as I bent to make the first incision.
>
> The doe's innards warmed my hands, and while I worked through the day, gutting her, hauling her home, skinning her carefully, butchering laboriously, and, finally tasting the grilled meat, I began to feel—corny as it may sound—redeemed. Part of what makes hunting such an intensely emotional experience is the physical responsibility you take for the death of your food.[5]

A counselor and former priest told me the story of his first deer: "I hunted squirrels and rabbits when I was a kid. Then I went off to college and got interested in the ministry. I forgot about hunting. After a few years, I decided that I wanted to work more with people as a healer than a priest and so I went back to school and got a degree in psychology. I was living around where I grew up, and one of my buddies from high school invited me to go deer hunting with him. The first morning of the season a four-point white-tailed buck stepped out in front of me. I couldn't believe how excited I suddenly became. I dropped him with one shot. My hands were shaking and trembling as I walked up to the fallen deer. At first I felt this wave of sadness, he was so beautiful, bigger than life. I was spellbound. Then, gradually, I thought about what had happened and came to realize I was finally being honest about killing the meat I ate. I knelt down and said a prayer. It was the most profound spiritual experience I have ever felt."

Emotional growth involves experiencing strong emotions and then learning to choose how to react to them, as well as coming to know that you can exert control over your emotions, for they are learned reactions. You can approach any life situation as a growth experience, staying grounded, feeling your emotions, and working through the associated thoughts until you feel clear. Hunting is no exception. Many hunters tell me they say prayers, silent and aloud, when they come upon what they just killed. Some cry, their tears a mixture of joy and awe. Seeing the death of an animal occur by your own hands makes one appreciate life even more. Taking pride in an animal you have shot is normal and healthy. It is proof that you have mastered one of life's most basic challenges. It affirms you and leads to your affirming of the animals. A sense of reciprocity grows, and in time love results. Love is the highest, most intense, most beautiful of emotions. It begins with self-love and then reaches out to the object that evokes such a powerful feeling. In time the two blend into one, yet keeping their own identities as well.

In the 1970s, I worked with the Esalen Institute Sport Center on studies of world-class athletes in many sports. Among the findings was that success is associated with unity of mind, body, and spirit. Typically, top male athletes have a strong anima. Intuition, emotion, and feeling are critically important to peak performance; the feminine side is the root of what makes them great. Legendary running back Jimmy Brown seldom got hurt because he could sense and feel other people as they hit him and make quick adjustments accordingly. Women athletes usually have a strong animus to express their aggressive energies, in addition to their strong ca-

pacity for feelings. Emotional intensity is essential for peak performance among all athletes, male or female. Intuition is crucial to
being at the right place at the right time, and spatial awareness en-
hances peripheral vision. Without the balancing elements of the
feminine, athletes could not excel.

There are some hunters who steel themselves up and play
macho roles. Their hunting psychology is more rooted in skills and
technique. They tend to be more tense in general, and probably
are prone to drink more on hunting trips to relax. They are missing out on an awful lot of fun.

Women Who Hunt

In Jean Auel's popular novel *The Clan of the Cave Bear,* the hero-
ine, Ayla, is banished from her tribe because she wants to hunt.
Looking cross-culturally, Emory University anthropology professor Melvin Konner concludes:

> With one impressive exception (the Agra of the Philip-
> pines where women hunt routinely and hunt well) all
> hunting and gathering societies on record have a division
> of labor by sex, with men doing most of the hunting.[6]

Modern society may be in the middle of a major transformation in sexual roles in regard to hunting. While the numbers of
hunters in general have been declining, the numbers and percent-
ages of women who hunt and fish in the United States are rising.
In 1992, of the 14 million people who bought hunting licenses in
the United States and took to the woods with gun and bow, 10
percent were women. Slightly less than 2 percent of all women are
hunters. (Approximately 16 percent of all American men hunt.)
More than 30 percent of the nation's fishing licenses sold in 1992
were sold to women. Typical of what is happening is in Michigan,
where 37,500 women hunted in 1984, and in 1993, 60,000 women
bought hunting licenses. In addition to sheer numbers, this is a

jump in women's share of the hunt from 5 percent to 7.5 percent of all hunters. Across the United States, the steady increase in women hunters is the brightest spot in hunting recruitment.

Among the leaders of recruiting women hunters are Christine Thomas, a University of Wisconsin Stevens Point professor, and Tammy Peterson who works with the United States Fish and Wildlife Service in Portland, Oregon. Together they have developed a workshop, Becoming An Outdoors-Woman, which has taught hundreds of women woods skills including rifle and shotgun shooting, archery, tracking, wildlife biology, muzzleloader shooting, camping, and backpacking. Since the first workshop in Wisconsin in 1990, the model has inspired programs in other states including Nebraska, Arkansas, Oregon, Montana, Washington, and Texas.[7]

Studies show that most women learn to hunt as adults with male friends or family members. Hunting traditionally has been seen as a male sport, but Thomas and Peterson say that this is changing and point out that more women becoming hunters is very important for the survival of hunting. One major reason is that, according to studies, 60 percent of the children born in 1989 will be raised by a single parent, usually a woman, for some period of time before the age of eighteen. Studies also show that if kids aren't exposed to hunting, they are less likely to become hunters as adults.

The Becoming An Outdoors-Woman workshop opens new doors for women, and it is helped by support groups such as Tomboy. Based in Oregon, the two-year-old Tomboy has seven hundred members and is growing.[8]

Another support group is the Wisconsin Sportswomen's Club, whose president, Nancy Putnam, began hunting at age twelve, because "we needed the meat." Putnam's choice of weapon for deer is a replica of a 150-year-old flintlock rifle. It allows her just one shot. "That's all you should need," she says. "Technology can separate people from the woodlore, from the life." She and her husband, Ed, camp in an eighteen-foot teepee and have learned traditional hunting and tracking skills of the Indians, as well as how to brain-tan the hide.[9]

Women hunters do seem to be more serious, as John O'Leary said, and they bring home the meat, too. Some other women, however, challenge not just hunting as being sexist, but eating meat itself.

Meat and the Feminine

In her book *The Sexual Politics of Meat: A Feminist Critical Theory*, Carol J. Adams theorizes that eating meat is an example of male dominance and sexual discrimination and advocates taking up "ethical vegetarianism" as a political statement about the treatment of women.[10] This theory follows the same lines as Merritt Clifton's charges about hunting being an example of male dominance that kills the feminine principle and exploits women.

To explore another view of this position, let's turn to Leslie Baer-Brown, an author and a nationally syndicated radio show host in Los Angeles, and a "recovering vegetarian." Leslie is deeply concerned about human rights issues and in 1991 she was a member of a team that flew into the Venezuelan jungles to make contact with the Yanomami Indians, one of the last surviving indigenous tribes of the Amazon rain forest, to mount a campaign to save them from extinction by logging and mining. Leslie and her team spent a week living in a Yanomami village, gathering information for a documentary video, *Yanomami: Keepers of the Flame*, which has since gone on to win many awards.

That one week with a rain forest tribe, living with people whose lifestyle has basically remained unchanged for thousands of years deeply touched Leslie. For a number of years she had studied with a spiritual teacher, meditated and practiced yoga, and been a strict vegetarian. Suddenly she was thrown into a situation where there were no health food stores, vitamins, or air conditioning. People hunted to survive. They killed monkeys, parrots, fish, snakes, and insects for food, as well as ate roots, nuts, fruit, and some simple agricultural crops. They were happy and joyful. Not wanting to change or criticize the Yanomami, which would have been damaging to their culture, Leslie ate as they did. While munching on macaw, she came to the realization of how "arrogant the whole moralistic modern vegetarian diet is. To keep healthy, it requires vitamins and other dietary supplements and it consumes enormous energy resources and causes considerable air pollution to transport foods long distances just to be politically correct. Vegetarianism is a luxury," she says, adding that living in cities "we have lost touch with our animal self. If we want to really get back into harmony with nature, we have to own our own identity and become a part of the food chain." Her remarks bring to mind Carl Jung's comment that "he who is rooted in the soil endures."

The jungle experience also made Baer-Brown recall a significant event in her earlier spiritual studies with an Eastern guru. One day she happened to walk into the guru's private office. He was smoking a cigarette and eating a hamburger. She was shocked. Both acts seemed inappropriate for a holy man. She challenged him, and he responded that he needed these simple vices to make him human. He then added "Is it not much worse, spiritually, to judge someone than to smoke and eat meat."

In ancient Greece, the worshipers of the god Dionysus were primarily women. They engaged in his worship by retreating to wild places where they joined together in *orgia* (the root word of *orgy*) ceremonies where passionate music from wind and percussion instruments drove them to wild dancing and trance states that led them into oneness with the god. In this state of possession in the orgia, the followers of Dionysus would attack animals and eat their raw flesh, often tearing it from the bodies of living animals. Modern translations of the spirit of Dionysus would seem to include the life of the late rock singer Jim Morrison and the cult of Charles Manson, although it should be noted that Manson's family was vegetarian in the name of ecology. Yet, the lives of these people say more about the dark side of this archetype than its potential power, for Dionysus is ultimately a healer of wounded female souls. In Greek mythology Dionysus is the only god who specializes in rescuing and nurturing women who have been dominated, even raped. The approach of Dionysus the healer is to evoke deep instinctual energies to recharge people with power, driving away repressed emotions and limiting beliefs. Following a Dionysian path, one constantly walks on the knife edge of violence, because of the powerful energies involved, but by learning how to serve a higher spiritual purpose one dances on the edge of the knife in ecstasy.

I introduce the Dionysian rites not so much to advocate them as to point out that what is "feminism" and what is being a powerful woman are different to different people, times, and places. The goddess has not always been "nice" historically. She stands for life and death. Sometimes she demands blood sacrifice.

"Vegetarianism is a long-accepted diet for spiritual clarity, physical health and potential longevity," wrote actor and martial arts master David Carradine in his immensely practical book *The Spirit of Shaolin*. However, based on many years of studying health and fitness in both Eastern and Western traditions, Carradine then goes on to say that diet should serve the purpose of making you healthy, and that the right diet will be influenced by your ancestry as well as your physical needs. When he needs to build up his body for a strenuous project, Carradine, who normally eats little red meat, may switch to a diet of raw red meat and water. To spice it up a little, he says he creates a body builder's snack or "Tarzan food" of "raw chunks of beef, marinated with olive oil, garlic, lemon juice and chopped onions."[11] If you eat a raw red meat diet for long, you will "start acting like a tiger around the house," Carradine cautions.

Diet does change consciousness. In conventional Chinese philosophy, meat is yang, which is masculine. Vegetables are yin, which is feminine. Following an all-meat diet will make you more yang. Following a vegetarian diet will make you more yin. This is one reason why people who are very nervous and tense sometimes benefit from a vegetarian diet, as it balances their yin and yang selves. For your health's sake, you should eat what you need to be healthy. If your diet denies your health needs—with or without meat—for whatever reasons, it will make you sick and, on some level, angry. If people who follow a vegetarian diet, which is supposed to make them more yin, spiritual, expansive, mellow and easygoing, are angry, moralistic, and self-righteous, my suspicion is that their diet is nutritionally deficient. The anger is an expression of the conscious/unconscious war going on within them, for at our innermost self, we do know what is best for us. Yes, you can choose an austere diet and live a pious and holy life. Yogis in caves in India do this. Some live to ripe old ages. The Sufis, a Middle Eastern spiritual sect, teach that there are seventy-two paths to God and they are all equal. One of these paths is the way of the hunter.

Hunting and the Feminine Principle

The women's rights movement has opened many doors and removed many barriers to women. Shifting away from a pure patriarchy has also broadened men's roles and raised the question to many: "What is it to be a man?" A popular response has been a men's movement, which is widespread and is having an influence on men's lives, in spite of criticism and a good deal of wisecracking from the popular media. When President Clinton went on his famous duck hunting trip at the Fruit Hill Farm on December 27, 1993, Kathy Tieder, the wife of the owner of the hunting lodge, told the *New York Times*, "It's the old camaraderie thing and male bonding bit." Ten years ago no one would have said anything about male bonding.

Men everywhere are going to men's groups, participating in drumming circles, buying books on men's psychology, and going on weekend retreats of sharing and ritual in droves, with the goal of discovering and reclaiming their masculinity. One of the reasons for this groundswell is that as women gain positions of power and status, men can let go a little and allow some of their emotions and feelings to come out, as they no longer expect themselves to be the sole breadwinners. The stoic macho hero figure, who shows no emotion and has steel nerves, is a man who is missing out on a lot of enjoyment in life and probably is contributing to his own demise in some kind of psychosomatic illness. The more feeling man and the more assertive woman are the role models for the future.

There is an assumption by some feminists, both female and male, that as this happens men will become more timid and hunting will therefore disappear. The data do not support this hypothesis. The softening of the modern male will likely cause a reduction of "slob hunters" who go out to drink, shoot, and play cards as a release and a relief from the pressures of life. But there will be counterbalancing forces among those who get more in touch with who they really are and do not merely kowtow to a stereotype to be successful. Psychoanalyst Marie-Louise von Franz observes that as societies take on a more matriarchal character, the "primitive side" of men comes out more fully.[12] The drumming circles, ritual making, and male bonding workshops are just the beginning of the men's movement, not the culmination. As men get in touch with more of their feminine side, their anima, some will hear the call of the hunting horn louder than ever. Entering hunting from this mind-set, we will see much greater interest in hunting with "primitive technologies" such as archery, atlatls, and muzzleloading rifles and shotguns. This is so because men will be hunting for the deeper experience of hunting as an instinctual, even spiritual act rather than to prove they can shoot the biggest trophy. Women who have a strong and healthy animus will join them. Some men will also create new hunting mystery schools; modern versions of the Order of Ogun. In Northern California there is at least one private hunting club that advertises optional drumming circles as part of a packaged hunt for hogs or deer. On hunts in foreign lands, such as Africa and Mongolia, native guides sometimes conduct traditional ceremonies to honor the hunters and the animals hunted, just as their ancestors did hundreds of years ago.

We will also see more and more women becoming hunters. In Pennsylvania, which has 1.1 million hunters in the woods every fall, 8 percent of the hunters are women, and the number is rising, which is the case across the nation.

In the 1960s we witnessed waves of social criticism from the collective social unconscious that became the civil rights, women's rights, anti-Vietnam war, and ecology movements. These collective energies challenged the dominant paradigm of the American dream and showed its inequities, including restrictions not just on women but on the feminine principle as a whole. As in an identity crisis of a person, the awareness that a problem exists in a society is just the beginning of solving the problem, not the final solution. If the animal rights movement does not polarize and become a full-fledged terrorist movement like the IRA, turning the hunting grounds into battlefields, it seems likely that the hunters of the future will return to being socially accepted heroes and heroines, for they will be hunting with both the masculine and feminine sides of themselves in greater balance, and, most important, because they themselves choose to hunt as an expression of who they really are.

Bracing against the pre-dawn chill in a duck blind with my father is one of the most vivid memories of growing up in Eastern Arkansas. Similar mental snapshots have been created across our nation for centuries as parents have introduced their children to the beauty of our wildlife through hunting and fishing.

CONGRESSWOMAN BLANCHE LAMBERT, ARKANSAS
Hunter and member of the House Merchant Marine and Fisheries Committee[13]

Eight

Hunting in the Future:
Will There Be Any
"Good New Days"?

When some of my friends have asked me
anxiously about their boys, whether they
should let them hunt, I have answered, yes—
remembering that it was one of the best parts
of my education—make them hunters, though
sportsmen only at first, if possible, mighty
hunters last, so that they shall not find game
large enough for them in this or any vegetable
wilderness—hunters as well as fishers of men.

HENRY DAVID THOREAU[1]

At the end of October 1993, I flew back to Michigan on business. After the meetings, I drove to Grosse Ile to pick up my father to take him to a football game in Ann Arbor. At ninety-two, he is the oldest living University of Michigan football player. He is more of a celebrity today than when he played on Field Yost's "point-a-minute" teams in the early 1920s. That night, lying in bed with the window open, as I had done all through my youth until I left for college, I listened to the sounds of the marsh. It was duck hunting season, and memories flooded back to me as I heard herons squawking, a raccoon chattering, the snort of a deer (we never had deer on Grosse Ile when I was growing up in the 1950s), feeding mallards quacking, and a dozen other familiar sounds of the night. Suddenly the marsh became silent. A fox barked and then there was a tumultuous honking of a flock of Canada geese. It has been well over a decade since I hunted ducks in the marshes of Lake Erie, but in those twenty years when I did hunt ducks there, I saw only one goose shot around Grosse Ile. They never landed or even came near the water in the old days. The next morning as the sun came up, I was awakened by the calls of the geese taking off, as a hunter's boat worked its way through the predawn darkness. Later, when I was having breakfast, two dozen mallards landed in the front yard and began to eat grain under the bird feeder. These were not the white Pekin ducks or the over-sized domestic mallards that are so fat they cannot fly—these were all legal mallards. My mother said that some days she has almost a hundred ducks, and even a few geese, come right into the front yard to eat cracked corn—even during the duck season!

I went out to check on the hunters. The blinds were in the same places. So were the ducks, the hunters told me. Not much had changed. More mallards and wood ducks, fewer black ducks and fewer canvasbacks, they reported. Not as many birds as there were fifteen years ago, but still plenty if you know what you're doing. What about the geese? I asked. Oh, they laughed. If you want to see geese, go up to the golf course.

The headwaters of Frenchman's Creek come from a spring that bubbles out of the ground in the center of the thirteenth hole of the Grosse Ile Golf and Country Club. A new subdivision had gone in just south of the course, and when I drove in, more than a hundred Canada geese were happily feeding in a pond created from what used to be a swamp where we speared carp when I was a teenager. The residents, it seemed, had been trying everything to get the geese out, because they were smelly and loud and people feared they would create a health hazard. Hunting is not allowed on the land of Grosse Ile, and firecrackers, smoke bombs, and even pet dogs had met with only limited success. Live trapping and relocation had worked for a few birds, but the adult geese just flew back, even when transported fifty miles away, the disgruntled club owners told me. The geese left only when the lake froze over, they said, and then came back soon in spring. My father and I drove up to the football game a little later. He doesn't get out hunting anymore, but watching the river as we drove along, he commented on how the bluebills seemed to be late this year.

The next day, my old friends Rusty and Craig wanted me to get my bow and go hunting with them. We packed up and drove down to Monroe to this special place they knew about. As soon as we turned the corner off Telegraph Road, we were greeted by two deer. We got out and got our equipment ready. In minutes Craig drew back and shot a buck right in the heart. Then Rusty bagged a caribou. A rabbit hopped out and I shot and hit him squarely. Then a moose, a bear, an elk, a mountain goat, a big-horned ram, a buffalo, and half a dozen more deer and elk walked right in front of us. We were shooting as fast as we could. I had never seen anything like it.

Our "hunting heaven" was the Dart Video Target System at The Archery Center of Monroe. At twenty-five paces we were shooting blunt arrows at wildlife videos projected onto a screen positioned two-thirds of the way down the hall. (Incidentally, the two deer on the lawn were decoys, not live. People now use life-size deer decoys.)

"Hunting" with a virtual reality system like the Dart is the most exciting and educational target shooting an archer can find. The targets are animals in wildlife films and television shows. Each is in its natural setting. As they come out onto the screen, which is twenty feet by twenty feet, one archer draws. You then have ten to twenty seconds to pick the best time for a shot and take it. When you shoot and hit the screen, the whole scene freezes, and while the arrow falls to the floor, you get feedback on your shot, which is marked on the screen. If you hit the animal in the sure kill zone, a red light comes on inside a white circle and you are given ten points. If you miss the bull's-eye but still hit the vital area where a quick kill is assured, you get five points. Outside the circle, you get no points. And you can't just shoot anytime. Just as in real life, you have to wait until you have the right angle and there is no brush in the way. If you shoot at the wrong time, the screen reads NO SHOT SITUATION and you get no points even though you may have hit the center of the kill zone on a difficult moving shot. As soon as you've gotten your feedback, the scene changes and another animal wanders out of the bushes and onto the screen. This is not just entertaining target practice, it is education. You quickly learn to wait for the right shot, the sure quick kill to the heart. This is hunter's education of the future. As virtual reality systems improve, more people will be "hunting" indoors.

That afternoon I went down to the Pointe Mouillee state wildlife area, about eight square miles of sprawling marsh at the mouth of the Raisin River, which has been an almost magnetic center for wildlife for centuries that in turn has drawn people. First an Indian encampment, then a haven for early hunters and trappers, in 1875 "the wet point" was bought up by eight wealthy duck hunters and became the exclusive Big Eight Shooting duck club. Then came the Depression, and with funds from the Pittman-Robertson Act, in 1945 Pointe Mouillee was purchased by the state of Michigan. Today, Pointe Mouillee is one of those places that duck hunters all around the world know about. Some say it is the pilgrimage place for duck hunters of the Midwest, for every September right before the duck season begins, more than thirty thousand people gather at Pointe Mouillee for the annual three-day Midwest Decoy Contest and Michigan Duck Hunters Tournament.

It was a crisp, clear October day with a strong west wind. At 3:30 P.M., when I pulled into the parking lot, one boat of three hunters in their early thirties was just coming in. They had killed their limit—four mallards, two redheads, one scaup, one ruddy duck, and one merganser.

It was the first hunt for one of the party, and they asked me to take their picture beside the boatload of decoys, holding up their ducks. They were truly happy people.

Out in the marsh you could hear an occasional shot, but closing time was not until 6:38 P.M. Overhead, clouds of starlings and red-winged blackbirds were swirling into the marsh and nearby fields to feed. Herons and egrets were everywhere. Three jacksnipe buzzed by on their way to safety. Terns plunged into the water just offshore, while fishermen on the bank enjoyed filling their buckets with the first fall run of yellow perch. An abundance of wildlife is a sign of a special place, shamans say. By this standard, Pointe Mouillee is clearly a place of power.

I went inside and talked with the ranger, who explained the regulations to me. The marsh is divided into three regions—one is a permanent sanctuary where no-hunting is permitted, another is an open zone where anyone can hunt according to state and federal laws, and the third area is for hunting by permit only. In the permit-only area, hunters are allowed to hunt only on Wednesdays and Sundays. There is a morning hunt and an afternoon hunt. Hunters are selected in two drawings, held the day of the hunt, one at 5:30 A.M. for the morning hunt, which goes until noon, and a second drawing at 11:00 A.M. for the afternoon hunt, which begins at 1:00 P.M. If you are drawn in the morning, you cannot hunt in the afternoon too in the permit area. A daily permit is three dollars.

Each lucky hunter is allowed to take only twenty-five shotgun shells loaded with steel shot no larger than BBB, which is roughly the shot size that air guns use. Winners are assigned to a zone, with a maximum of three people in a party, and cannot hunt in any other zone. Alcohol consumption is not permitted, no cans or bottles may be brought into the marsh, and no permanent blinds may be built. Also prohibited are outboard motors over 10 horsepower, fires, and shooting from any road, dike, or trail. People who do not win in the draw may hunt in other areas of the marsh, or venture out into Lake Erie, where any place is legal. On the first Sunday of the season, priority in the afternoon drawing is given to junior hunters, twelve to sixteen years old, accompanied by parents or guardians.

These restrictions may seem harsh, especially to old-timers. But the permit-only area almost always fills up quickly, the ranger said. Hunters had been averaging 1.7 birds apiece to date. Not bad for a half-day's shooting when the limit per day is only three, and you are allowed to take no more than one black duck, pintail, redhead, hen mallard, and no canvasbacks in the daily bag limit.

While I was waiting for the day to end so that I could see the hunters coming in, I walked out on a mile-long earthen dike that marks the sanctuary area. Within gunshot range of hunters staked out on the hunting-zone side of the dike were hundreds of mallards, wood ducks, widgeon, and scaup happily feeding in several ponds in the no-hunting area. Countless others moved in and out of the thick tangle of willows, bulrushes, cattails, and phragmites along the edges. The ducks knew which areas were hunting areas and which were not! Half an hour before closing, ducks were everywhere in the sky, coming in to feed for the night. Across the marsh, duck calls poured out invitations, and shooting was furious. Then, at 6:38 P.M. gunfire stopped, and within minutes what had looked like pieces of the marsh revealed themselves as boats as they headed back in to the ramp. Each hunter with a permit must check in, and so I waited inside as they brought in their kill and reported whether they had shot any ducks that they could not find.

One of the first hunters to check in was a fifteen-year-old boy who came walking across the parking lot carrying a Canada goose he had shot. He was so excited his feet barely touched the ground. After he had checked in I asked the ranger about the geese. It turned out that it was the first one that anyone had shot that year. "We've got our own flock," the ranger said. "Every year we get calls from people around the Detroit area with problem geese in their ponds. We trap about three hundred or so and bring them down here to release them. The adults fly right back, but the juveniles sometimes stay. Costs a lot of time and money to try to control the geese, and it really doesn't work too well."

The hunters straggled in, all men between fifteen and fifty-five, wearing camouflage clothing, some with faces painted, most wearing hip boots; outside, their cars and trucks pulled trailers with special duck boats piled high with decoys. In the cars were tired dogs, noses pressed against windows, who wanted to go home for dinner. The average bag that night was about three ducks per zone, mostly mallards. The men all knew the sex and species of the ducks, and out of thirteen parties that I watched check in, there were four ducks reported downed but lost. Bill Gilbert once wrote in *The Saturday Evening Post* that:

> Hunters are noisy, belligerent, and the dirtiest of all outdoor-users, littering the landscape with bottles, corn plasters and aspirin tins. They are also dangerous. . . . Sports hunters are, as a rule, disreputable, the most obvious complaint against them is that they are destructive of wildlife. . . . All hunters must be skinned of the right to use the forests and fields as if they were a personal preserve, a private butcher shop.[2]

I saw none of those qualities in the hunters who checked in at Pointe Mouillee. Perhaps we have a new species of hunter. I think we do, but the hunters described by Gilbert have never been common.

That night I went over to Rusty's, my bowhunting friend, for dinner. His mother, Mechel, cooked venison steaks for the special occasion. It was a chance to tell stories, recapture some memories, and start up the fires for another year. Time to bring out the guns, check out the new ones, and remember the old days. Piecing together archery, rifle, and muzzleloader seasons in Michigan, you can hunt for nearly three months, and kill three deer. Still the herd is growing, and crop damage and auto-deer collisions are increasing. Many people build their whole year around deer season. Others spend their life savings on a deer cabin and a piece of land in northern Michigan. The people who have the best deals in hunting are those who either own or live in rural areas adjacent to public lands. Urban hunters pay a price both in money and in access to lands. Probably the best thing that has happened to urban hunters is the invention of virtual reality target ranges. I predict they will become more popular than bowling and pool in some places. I have to add that when I happened to talk with my hunting buddies back in Michigan in December, about half had gotten deer. In Michigan, whether you got your deer for the year is still acceptable conversation. Rusty's sister, Ann, and her husband told me that while driving five hundred miles north to their deer camp they had counted three hundred deer coming back on tops of cars. In some parts of the United States, hunters are still heroes and heroines. Then there is California.

Hunting in the Golden State

California is a land of contrasts and extremes: snowcapped mountains and Death Valley, rain forests in the northwest and arid deserts in the southeast; records set for coldest temperatures in the nation at Mt. Whitney and warmest temperatures at Needles; plus volcanoes, hotsprings, and earthquakes. We have densely populated urban areas and ghost towns. By 2010, white people will be a minority. A recent tally of animal rights groups in the United States listed 235. California is home to 56 of them. The ratio of antihunters to hunters is probably higher in California than any other state. The U.S. Justice Department reports that 54 percent of the animal rights terrorist incidents in the nation have taken place in California. You want it, we've got it.

Turning east off Highway 101 just north of Santa Rosa, you wind up into the rugged Sonoma mountains, which become increasingly arid and desertlike as the road snakes eastward. Then, just as you think you are about to enter into a desert with cacti, roadrunners, and lizards, the land drops away in front of you and you descend into the fertile Sacramento Valley, with a striking backdrop of the snowcapped Sierras to the east.

Farms appear, first wheat and vegetables irrigated by sprinkling systems. Descending lower, irrigation canals cross the road, providing water for flooding rice fields. Turning north, toward Mount Shasta, I take I-5 to the tiny farming town of Maxwell, where I stop at a general store and gas station for food and directions. I ask the clerk about the ducks and geese. His face lights up as he says, "Oh, great year, lots of birds down!" He points to the road I need to take to get to the Delevan National Wildlife Refuge, one of a string of wetland preserves in the Sacramento Valley.

The landscape is flat. I scan the air for life, but I can see only seagulls at first. It is midafternoon, not a good time for flights of ducks and geese, especially when it is a crisp, clear December day. I turn onto a dirt road at the sign for the refuge and immediately am greeted by two cock pheasant in the middle of the road. Hunting season for pheasant has just closed, and the birds know it.

Suddenly, I see ducks and geese in the air, off to the east. A quarter of a mile farther I see half a dozen flocks crisscrossing overhead. The road now cuts through marshland, and I see ducks and coots swimming in pockets of open water, while wading herons and egrets stalk nearby. More birds are in the air. I come to a trailer, which is the hunter check-in station. In the parking lot are five pheasant pecking in the gravel like barnyard chickens. I pull in and shut off the motor, ending the guitar wailing away on the radio. Normally, when you stop after a long drive, your ears ache with the stillness. Here, as I open the car door, I am greeted by a din. All around me from the marsh is a swelling chorus of honking, quacking, and whistling. I get into the bed of my truck to look above the marsh grasses, and immediately to the north I see a white blanket of at least fifty thousand snow geese flocked together, some within thirty yards of the road. Many are standing in tight groups on mounds rising out of the marsh, and in the water beside them thousands of darker ducks of half a dozen species are busy feeding. Here and there among the birds are small flocks of chunky, grayish brown specklebelly or white-fronted geese. Valhalla, they say, is a hunting ground where wildlife is teeming. I have just arrived in heaven.

It is Tuesday in mid December. Today there is no shooting, as hunting is allowed only on Wednesdays, Saturdays, and Sundays, and only in certain parts of the refuge. Tomorrow there will be hunters manning forty-six blinds built on raised mounds scattered throughout the marsh to the south of the road. The area to the north of the road is one of the no-hunting areas, and the ranger at the check-in station tells me that the ducks and geese seem very aware of the difference between the south and north sides of the road, at least most of them. She says that if I want to see even more geese and ducks I should head north to the Sacramento National Wildlife Refuge headquarters, as there is a slightly elevated road there that will take me right out into the middle of the marsh to watch the birds. I take her advice and soon am driving through a goose and duck metropolis of close to half a million waterfowl, as well as numerous pheasant that almost have to be shooed out of the road. Taking this tour is free, a gift to anyone who wants to see our web-footed relations, donated by hunters from Pittman-Robertson funds.

Only one other time have I been in such a teeming mass of ducks and geese. That was a snowy December day along the Columbia River at the eastern Washington and Oregon border when at times it seemed that mallards and Canada geese blanketed the river from bank to bank. That too was a refuge area. Considering that California has lost nearly 90 percent of its original wetlands, one wonders what the Sacramento Valley would have been like when the first white men dropped down out of the Sierras; it must have been a marshland of enormous proportions, interrupted only by the strange Sutter Buttes, which suddenly rise, as if to watch over the spectacle of the wintering flock of the Pacific Coast flyway as they rest and feed. The birds have lifted my spirit. It is regenerative just being in their presence, a far cry from the draining feeling I get when I am in most cities.

Hunting at all four of the refuges in the Sacramento National Wildlife complex—Sacramento, Colusa, Sutter, and Delevan—is tightly controlled. A maximum of only 601 hunters, total, can hunt here for waterfowl on an open day—Wednesdays, Saturdays, and Sundays, during the open season. The Wednesday after Christmas is a special day for junior hunters only. The demand is high for these places, and to give everyone a chance the state runs a lottery. To get into the pool, you go to a sporting goods store, buy an orange computer punch card, punch out the date and refuge you want to hunt, write down your name, address, and hunting license number on the card, and then buy a $1.05 wildlife stamp, which you put on the computer card. You then mail the card to Fish and Game in Sacramento at least ten days before the date you want to hunt. You can apply for only one refuge on any given day. If you get selected, you can bring three friends, but you all must hunt from the same blind. You must first buy a hunting license and a federal and state duck stamp just to apply.

If you are lucky enough to be selected in the lottery, you must arrive about two hours ahead of shooting time, which means about 4:00 A.M., and then you will be assigned a blind, after which you buy a one-day hunting permit for ten dollars a person. Steel shot only, no shotguns larger than a 12-gauge, and each hunter may carry no more than twenty-five shells into the field. Once you reach the blind and set out the decoys, you have to get out your field guide to make sure you are shooting at legal birds. The limit is four per day, eight in possession. In the daily bag you can have no more than three mallards, only one of which is a female, only one pintail of either sex, and no more than two redheads and/or canvasbacks. On these refuges you cannot shoot at Canada geese, which are not that abundant here.

For the 1993–94 duck season, I accepted the regulations, taking the time to study the pile of guides for state and federal waterfowl hunting. Way back in October I paid my $49.90 for my license and stamps, and then this year my son and I applied to hunt at Delevan for every Saturday and Sunday of the season, as well as two Wednesdays during Christmas break. No luck. We were not drawn. Some of the free-roaming areas on the refuge are determined by a lottery drawn the day before, but it takes two and a half hours to drive from where I live to the Sacramento Valley. Do you drive up the day before with all your gear and hope to get drawn? Or maybe you drive up, get in the drawing, and then have to go home and get everything and drive back up to a motel—seven and a half hours on the road in one day—so you can be there at 4:00 A.M. the next day? If you aren't there on time and you've got a number, you lose your place. If I haven't got all these regulations exactly correct, I apologize. I only have a Ph.D. in natural resources, and I have a little trouble reading eight-point condensed-type manuals of regulations.

Later, after the hunting season closed for the year and I read the tallies for the federal refuges, I felt affirmed by my intuitive pull to choose Delevan as a place to apply for duck hunting. The 5,553 hunters who were drawn and hunted Delevan shot an average of 2.05 waterfowl per gunner—10,072 ducks, 1,274 geese, and 31 coots—which placed it among the top three public refuges in the state in 1993.

In a place with so much wildlife, located so close to major urban areas, you have to regulate hunting. There is no choice. It has been done this way for years and the ducks and geese keep coming back. At the time of this writing, however, there are mounting challenges to hunting on National Wildlife Refuge lands. It is possible that someday hunting on National Wildlife Refuges could be made illegal.

There are some other options if you want to hunt the Sacramento Valley. Just a few miles down the road is a hunting club where you can pay $65 to join nine other hunters in a single sunken-pit blind for half a day. Other daily rates for guided hunts run about $150 a person for smaller blinds. Then there are the more exclusive duck clubs. Just like country clubs, membership can run anywhere from $4,000 to $5,000 a year, plus an initiation fee, up to $1 million for the really plush places with full hotel accommodations, personal guides, dogs, and private heated blinds. Where there are no refuges or private duck clubs in the Sacramento Valley, there are farms, including some with flooded rice fields. Just like anyplace else, it helps to know people in the right places, and believe me, these people have been asked, and asked, and asked for permission to hunt. Few say yes to anyone but longtime friends. And let's not forget the freeways and other aspects urbanization that are mushrooming just a few miles south of the state capital, Sacramento, which is directly under the flight pattern of millions of migrating ducks, geese, sandhill cranes, shorebirds, and swans.

In the last decade California's hunting license sales have declined by one-third. Like other states across the nation that are losing hunters from their ranks, chief complaints of dropout hunters include complicated and confusing hunting regulations, shortened seasons, reduced bag limits, and the scarcity of public lands. You can still hunt if you live in California, but for many Californians when they set foot in the woods to hunt, it is not in this state. On some days there are more California hunters in Colorado, Montana, Nevada, Oregon, and Idaho than stalking the woods and marshes of California. Bear and caribou hunting draw many to Alaska or northern Canada. Hunting elk and native wild sheep in Siberia is becoming an "in" thing to try. Goose shooting is fabulous in Texas. If you have the time and money, Argentina is a new mecca for hunters, offering duck and goose shooting that is like what may have existed in the United States two hundred years ago. Africa is always a possibility, time and money permitting. Safaris, apparently, are becoming popular again.

Hunting isn't what it used to be in California, but there are some bright spots. One is the growth of private hunting preserves stocked with human-raised native and nonnative species. Some antihunters are outraged at hunters hunting pen-raised pheasant, ducks and quail, and deplore big game ranches with stocked buffalo, Asian deer, goats, sheep, and even African antelope. There are at least four thousand game farm/hunting preserves across the United States, one thousand of which are in Texas. Some are specifically managed for breeding big game trophies. At the Aspen Ridge Resort in Oregon, the price of the hunt includes an old-fashioned Sharps rifle and a ride out to hunt buffalo in a horse-drawn wagon. Others offer deals where you don't pay for a hunt unless you are successful.

Shooting preserves offer longer seasons and a much greater chance to experience success in hunting, which are important to new hunters and to keeping more seasoned hunters from hanging up their guns and waders. I've tried hunting on private shooting preserves, and it isn't the same as the wild, but to me, in addition to what it does for preserving hunting, it seems a better deal for the animals than the sure fate that awaits a domesticated cow, chicken, or pig that lives its entire, short life within fences and without even a taste of the wild.

And speaking of pigs, feral pig hunting is a bright spot on the California hunters' menu. The season is open year-round, the population in 1993 is approaching one hundred thousand and multiplying like rabbits, much to the dismay of farmers and park managers, because pigs don't just eat grass and acorns like deer do, they love to root in the ground like Rototillers, causing mud slides.

Hunting around our house is a ritual act. We try to get out whenever possible, and prior to Thanksgiving and Christmas we make a determined effort to hunt for the festive meal. I already told you about the goose we didn't get for Thanksgiving. While we didn't get drawn for waterfowl hunting in the Sacramento Valley for a Christmas goose, we were lucky enough to be picked from the lottery pool to hunt for wild pigs at the Lake Sonoma Recreation Area just south of Cloverdale in Northern California. Lake Sonoma is man-made, created by the Warm Springs Dam, which was built by the U.S. Army Corps of Engineers for water storage for irrigation, flood control, and recreation. The lake has two arms, one nine miles long and the other four miles long, caused by water backing up into the valleys of two creeks. Except for the vicinity of the dam, the land along most of the lake is steep hillsides with open meadows and densely forested thickets. Not good for any human use, but fabulous if you are a deer, turkey, or pig.

We drove up the day before to scout out the area. All hunting at Lake Sonoma, for deer, turkey, and pig, is done by lottery selection, which allows a maximum of forty bowhunters or thirty shotgunners per day, which is either a Tuesday or a Wednesday. It was late December, but it was near seventy degrees and sunny. We drove up to the Liberty Glen Campground and parked the truck. Within a hundred feet of the parking lot we saw fresh tracks. Soon we came upon a hillside that looked like it had just been plowed for planting. On closer inspection, the digging had been done by cloven hooves of pigs looking for acorns, grubs, mice, and whatever else they could find. Signs of pigs were everywhere!

We drove into Cloverdale to check in to a motel. The woman who ran the hotel asked why we wanted a wake-up call at five o'clock the next morning. "Hunting at Lake Sonoma," I said quietly, worrying that if she was antihunting we would be looking for another place to stay.

"Oh, pigs!" she exclaimed with a smile, "Good luck." Friendly natives.

The next morning we drove through a thick blanket of tule fog, which felt like light rain, as we meandered up the road to the check-in site. As we crossed the bridge at the lower end of the lake near the dam, the sky was suddenly crystal clear, and it was twenty degrees colder. Puddles along the side of the road were frozen. There were already a dozen cars waiting to be checked in by the Fish and Game warden. Some had boats to reach more remote areas, others were going to walk. As the eastern sky grew light, we found ourselves perched on top of a hill looking down onto a sea of silvery fog that went in all directions as far as the eye could see.

We drove to a nearby parking lot and got our bows and arrows ready. All hunters were archers on this date. I reached into my backpack and pulled out a small bag of cornmeal. I offered a handful to the rising sun, then to each of the four directions, the sky above, and the earth below, tossing small amounts to each. Then I took out a second handful and scattered it in the bushes slowly, talking to the pigs, telling them why we were there and what we planned to do. I told them we would take only close shots, shoot to kill, and wanted only one for Christmas dinner.

Then my son and I stalked off into the fog, keeping about fifty yards apart. We worked our way slowly down the hill along a ridge top, checking out the steep ravines on either side. There was excitement in the air. Off to the east, down in the fog we heard some pigs grunt. A few minutes later some wild turkeys gobbled across the lake. I peered over a ridge and fifty yards away was a four-point black-tailed buck. The deer season was over, he was safe, and he knew it, as he meandered off, flicking his tail as if to taunt us.

We worked our way down to the lake, and as we got to the last hundred yards or so, the pig sign disappeared. The grassy hillsides were undisturbed and there were no tracks in the soft earth, except those made by deer. This meant that the pigs were feeding almost exclusively on acorns, which grow on oak trees, which don't like wet soil. The night before I had a dream that showed a map of the lake with the higher elevations circled. The dream seemed to be an accurate account of where the pigs were. Dreams like this are nature's way of saying thanks for showing respect for the animals you hunt, I have been told by my Lummi friend Kenny Cooper.

We moved over to a wooded ravine and started to work our way back up toward the top of the ridge. Suddenly my son points. A shape moves in the bushes. A hunter in camouflage clothing has already started to do what we had in mind. Seeing that he was working back up the hill, we went back to the ridge top and began to climb, hoping to get ahead of him and let him drive anything in the ravine to us.

A quarter of a mile up the ravine, we met two other hunters, sitting down to catch their breath. Like us, they had heard pigs but had not seen any. It was 10:00 A.M. by now and getting warm, as the tule fog was long gone. Several times we saw large black-tailed deer, including three big bucks, sneak off in the bushes as we got within a hundred yards. Despite regular hunting, deer were everywhere, and they had impressive antlers.

We split off and started slowly to pick our way down the side of the ravine into the thick forest. Pig sign was everywhere. We moved down about a hundred yards and then I heard a branch crack. We froze. A dark shape moved below us. It was a large black razorback hog, working his way up the ravine ahead of the hunter who was walking the creek bottom.

Adrenaline started to pump. He was coming closer. I could clearly see the pig and judged it would weigh close to two hundred pounds. A window opened between the trees in the creek bottom, and there he was, fifty yards away, down a very steep hillside. I raised my bow to draw, and then he moved. No chance for a shot.

Another minute later, the hunter in the creek bottom appeared. We acknowledged each other and I said softly that the pig was ahead. His partner was on the other side of the creek, he said, so all four of us lined up and began slowly to work our way up the ravine, about fifty yards apart. We walked almost to the top of the hill. No pig. He had either crawled into a hole and watched us walk by, or climbed straight up a steep open hill. Most likely he laughed at us as we stalked past him. You don't get to be that big in a place where you are hunted regularly unless you are awfully smart.

It was now noon. I had to get back to the city, so we had to quit. Back at the parking lot, we met a hunter who had just shot a smaller hog and was getting ready to go back down the hill to drag it up. At the checkpoint, the Fish and Game man said that one pig had been checked in so far. A friend we met on the road had seen a herd of fifteen, but had not gotten a shot.

That was it. No second chance. The pigs would be hunted until 3:00 P.M. and then would have a week off in that area, when another lottery draw would be held. No fresh pork loin for us that night, but no encounters with antihunters either. Previously they had selected a junior deer-hunt day to harass, making sure that the kids were a target of their verbal insults. From what I hear, it was lucky that no shots were fired by some of the upset fathers.

We could have paid anywhere from $150 to $300 each to hunt for wild pigs on a large private ranch nearby, having the whole place to ourselves for the day. When we've done this, it is like stepping back in time, monster black-tailed bucks watching us try to catch up with bands of roving wild pigs. No other hunters. Complete silence, except for the gurgling croak of ravens and coveys of quail whistling in the blackberry thickets along the banks of the creeks.

We did have wild boar for Christmas dinner. Polarica wild game meats in San Francisco offers wild boar loins for $11.95 a pound. It was good. Very little fat. Not at all like domestic pork. Such is hunting in California. Limited land open to hunting near big cities, lots of regulations, and lotteries. The spirit of the hunt is still kept alive, but the old days of freedom to hunt whenever and wherever you want will never come back unless you are extremely wealthy or have friends in the right places.

Hunting in California is tightly controlled. Bear hunting, for example, is closely regulated and serves as an example of what is possible with modern wildlife biology. State-of-the-art wildlife census research estimates that there are between 17,000 and 24,000 black bears in California, the highest estimated population in a decade. The state sells up to 15,000 bear hunting licenses for $22.50 for residents, $145.75 for nonresidents. The 1993 season was set from October 6 through December 26 in a designated region of the state. About half the state is closed to all bear hunting.

Game laws require that each hunter send in his or her tag immediately when a bear is killed. An annual quota is set for the hunting season, which in 1993 was 1,250 (sows with cubs cannot be killed). Hunting was so good that year that the season was closed six days early, the closure being announced in all newspapers around the state.

California hunting seasons are often short and very complicated due to the state being divided up into many zones, each with its own special regulations. Licenses are expensive, and regulations are confusing; we have so many stamps, tags, and lotteries that you need a computer to figure out what to do. All this is further complicated by antihunting groups and a diverse multiethnic racial mixture with many people who do not hunt or who come from parts of the world where there are no game laws. More deer are shot here by poachers than by legal hunters. Every week or two you read in the paper about some new immigrants caught hunting squirrels in a park or fishing with illegal nets or lines.

The summary for our 1993 hunting season is a total take of ten ducks, all mallards shot at a private hunting club. Total spent for hunting was about $1,500. If I were still living back on Grosse Ile, I suspect I would have spent about $250 and had fifty ducks and maybe a goose or two to show for it, and maybe a deer taken at a friend's cabin up north.

Some people say that California is the forerunner for the whole United States—the place where trends start. Right now, if California is a vision of what is to come for hunting, the future seems an increasingly bad dream, an example of why researchers at Cornell University have predicted that hunting as we have known it may be gone in fifty years. Hunting is challenged now and in some places things do not look good, but the future of hunting is not an unstoppable nightmare, even in California.

I'VE CHOSEN THESE THREE STORIES about hunting because the settings are wildlife refuges close to major metropolitan areas. I believe places such as Pointe Mouillee, Lake Sonoma, and the Sacramento Valley National Wildlife Refuges are key to the future of hunting and to ecological awareness in general. In an age where information overload, stress, boredom, and crime run rampant through urban minds, refuges near urban areas are sanctuaries, not just for the animals—they are sanctuaries for people. They are places of the heart for man and beast, islands of sanity in a world that often seems crazy. Because they are so heavily regulated, when you do get a chance to hunt there you know it will be a quality experience. And if you just want to come and hope to see some wild animals, for most of the year the places are open and wildlife is abundant.

If we can preserve places like these refuges and let them be meccas for the wild animals, as well as for hunters and watchers, they will set a baseline that will spread out to other wilderness areas. These sanctuaries will stand there to remind us of our hunting heritage, and keep us honest about the preciousness of life. *Homo sapiens* living in the United States consume about 7 billion pounds of fish, 36 million cows, 5 million sheep, 80 million pigs, 380 million chickens, and 240 million turkeys a year. We see the animals alive on farms, and then, magically, they are at the supermarket. The vital link in between is a secret viewed only by a handful. If we all could even witness what must be done to eat meat, we would be a much more reverent people.

Those of us who need and want meat could not all hunt for food. Population densities for hunting and gathering cultures are seldom greater than ten square miles of wild land per person, and that kind of open space is just not there for most people anymore. Domestication of animals is necessary to feed meat to all the people who want it. If you choose not to eat meat, for whatever reason, then consider the added richness of life you would experience from harvesting even just a few wild vegetables, fruits, or nuts for your table. We need to nourish ourselves with meaning as much as with food, especially in a modern world where meaning seems continually to be set aside in the name of convenience, progress, and conformity. Social trends and fads sweep across America today because we hunger for identity, and to be up-to-date uplifts our self-esteem, briefly giving us a chance to stand out from the crowd. In the act of hunting, we rekindle what Carl Jung called our "ancestral soul," which is that primal part of us shared with all human history. In those moments when the spirit of the hunt possesses us we remember who and what we are, have always been, and will be for thousands of years to come. It will take a long, long time for the hunting instinct in man to disappear, if ever.

More and more hunters are realizing the importance of conservation to the experience of hunting, as across the nation the growth areas of hunting are in "primitive technologies," such as archery, muzzleloading black-powder rifles, and even the atlatl spear-throwers. Studies show that as many as 33 percent of all deer hunters now are taking up archery. Technology today gives hunters more potency to kill than ever before, but hunters are voting to go back to methods that place more of a handicap on them. This trend underscores that for many, the goal of hunting lies in the heights of the personal experience of the hunt, as much as or more than just putting meat on the table. The place of hunting in modern technological society is ultimately as a ritual act. When we can approach hunting with reverence, acknowledging the myths, symbols, and magic of the hunt that link us back to the Olduvai Gorge, our lives will take on more meaning, and we will know more peace through honest self-acceptance.

As for the lotteries, hunting itself is a gamble and always has been. I don't mind taking my chances on being drawn if I have to. If you go on a hunt and have a good experience, even if you don't shoot anything, the memories are the best thing you can take away. Hunters these days ultimately hunt memories as much as meat to put on the table. Memories feed dreams, and hunters must have dreams to keep them motivated. When you lose your dreams, you lose your mind. Hunts that once were common events for a majority of the population have become for many once-in-a-lifetime experiences. The scarcity of hunting makes it more precious. In the old days, hunters made sacrifices to the gods before venturing out into the wild. Today, hunting itself has become more of an act of pilgrimage, paying homage to a god whose presence in modern life has become less apparent but whose power has not diminished.

THE CAMPFIRE HAS BURNED DOWN to glowing embers now. In the sky, the watery face of the hunter's moon has again become the principal illuminating force, sending a ghostly luminescence across the land, making spiritual forces more apparent.

There is an old Chinese saying that birds are entangled by their feet and men by their tongues. The spirit of the hunt is here. I sense its presence all around us as the darkness grows. The ancestors are watching. Maybe they will join Artemis or Ogun or one of their allies and startle us in our dreams, where past, present, and future all blend into one and occasionally surprise us with glimpses of truth and awe. As we retire in search of new dreams to keep the fires in the heart aglow, let's let a wise voice speak last. I can think of none better than mythologist Joseph Campbell who said to me one evening over dinner in a French restaurant that the trouble with society today is that people have forgotten the basic law of life that "flesh eats flesh."

The planter's view is based on a sense of group participation; the hunter's, on that sense of an immortal inhabitant within the individual which is announced in every mystical tradition, and which has been one of the chief tasks of ontology to rationalize and define. The two views are complimentary and mutually exclusive, and in their higher stages of development, in the higher religions, have yielded radically contrary views of the destiny and righteousness of man on earth.

JOSEPH CAMPBELL[3]

Appendix

Suggestions for Saving Hunting

All hunters aren't automatically "bad guys," though some people like to portray them that way. Yet even if antihunting sentiments were to decrease, hunting would still be in trouble. The reduction of animal populations due to poaching, and of animal habitats due to the increase in human population and urbanization contribute to this problem, as well as the lessening of leisure time for hunters. The next decade is crucial for the future of hunting, and I offer some suggestions for how to help save the sport.

Join groups and form coalitions. There's safety in numbers. Animal rights activists get a lot of press, but this is deceptive: There are fewer activists on the membership rosters of animal rights groups than there are licensed hunters. If all hunters united for the cause, the sheer size of the group would cause the animal rights movement to lose power and influence. Alliances could be formed with other people who work with animals—pet breeders, farmers, circuses, animal trainers, and animal researchers. In January 1994 Idaho Fish and Game teleconferenced a "town meeting" for hunters around the state. Each state could do this.

Don't engage in revenge tactics against antihunters. Assuming an "eye for an eye" attitude will destroy hunting faster than any animal rights group action could. Most polls show that about half of the United States population supports sport hunting and half does not, with the least amount of support in major urban areas and on the West Coast. However, polls also show that a majority of Americans do not approve of animal rights activists who destroy property or who threaten or harm hunters. It is essential for hunters to have debates with animal rights groups; many of the members are well meaning, and should not be stereotyped any more than hunters should be. Mirroring the actions of the antihunters—targeting them with hate mail, threats, and media sensationalism—will only breed hatred in the hunting organizations and turn public opinion more against hunters. This negative posture might also attract new leaders to the hunting organizations, leaders who care less about hunting, animals, and the environment than they do about their own power. The bottom line for hunting is to avoid becoming an antianimal rights movement.

Hunting organizations need to build a positive image of hunting, as well as provide opportunities for both kids and adults to get into nature and learn about hunting and fishing.

Create a new image for hunters. Outdoor-sports journalists, rather than merely writing about techniques, locations, and equipment, could help the general population to think critically about hunting issues. These journalists could teach people to think ecologically, could develop feature stories on the good work being done by hunting organizations, and could profile celebrities who hunt. They could write about the animal populations and habitats that have been restored by hunting associations.

Some reporters have a predefined bias against hunting, and outdoor-sports journalists need to debate the hunter-bashing position these reporters assume. For example, an Associated Press article of late January 1994 that ran in the *Marin Independent Journal* carried the headline, "Fewer Birds Worldwide Shows Earth In Danger." The first sentence of the story—which cites a World Watch Institute report—begins, "Hunters, farmers, predators and gourmets are posing an increasing danger to some of the world's wild birds." Later in the article it is clarified that the hunters of the headline are Southeast Asian poachers and market hunters as well as subsistence hunters in Third World Countries. Readers might not have immediately understood that the article was discussing a small subgroup of hunters. The tendency to lump all hunters—good and bad—together has the unfortunate effect of stigmatizing all hunters as a group. The media should concentrate on an image of hunters as lovers of nature, who work to support of the environment, which is more accurate though less sensational.

Help create a legal task force to protect hunting. Hunting groups need to build a strong defense system to protect themselves from the challenges of the animal rights groups, which attempt to stop hunting through electoral and legal action. Assume they will do anything—grandstanding, misrepresentation, backstabbing, loopholing, litigation threats, and so on—to cause problems. Hunters need to find ways to counter these tactics, which not only push fish and game departments into a corner, but, more importantly, siphon off millions of dollars in wildlife management funds to the justification of hunting rather than to the actual managment of wildlife. The hunting community needs to take the offensive. It needs to secure hunting as a legal activity before the sport, and age-old heritage, is destroyed by the emotions of the moment, flawed by misrepresentation.

Organize prohunting community activities. Outdoor-sports and boating shows are now a primary way for people to learn about hunting and hunting organizations while they have fun. In addition to these shows, we need more hunting-related festivals, fairs, and art shows such as the duck-hunting and decoy festival held every September at Pointe Mouillee.

Promote community services. The hunting ethic, since the Paleolithic age, has included sharing the meat from the kill. When I was growing up in Michigan, local churches depended on donations of meat from hunters for the annual wild-game dinner fundraiser. The Sportsmen Against Hunger campaign, promoted by Safari Club International and a host of other hunting organizations, is one of the best recent developments for hunting. Hunters should continue to express the hunter's compassionate spirit, which is as old as hunting itself, through similar projects.

The programs to feed the hungry with donated wild-game meat have grown too large to be dismissed as a "publicity stunt," as some antihunters charge.

Support the men's movement and women hunters—either or both may help save hunting. It is too early in the men's movement to anticipate its future. However, I've read articles in men's movement magazines about men who've decided to try hunting because it's an activity in which men have participated throughout history. The men I know who came to the sport this way are among the most ethical of hunters, clearly falling into the nature hunter category. They are the new type of hunter that hunting needs.

Women hunters not only swell the ranks of hunting, they bring more children to the sport. The presence of women can also tone down rowdiness, and, I think, elevate the sport to new ethical levels.

Create more opportunities for hunting that don't require large cash outlays. In 1992, California raised $150,000 by letting non-profit organizations auction off hard-to-get tags for big game—three elk, one bighorn sheep, and ten antelope. Although this is a good idea, it is also an example of how hunting is becoming a sport that favors the wealthy. Rather than auctioning the expensive tabs, they could be used as rewards for people who do environment-related community service such as planting habitats, making nesting boxes, taking kids on field trips, and so on. We need to find more incentives for people to work for the environment.

Consider dropping the word "sport" to refer to hunting. Fish and game departments are quietly changing the vocabulary of hunting; for instance, they are replacing an emotional word such as *kill* with *harvest*. The word *sport* has many negative connotations when it is used to describe hunting. I have heard that the California Department of Fish and Game is phasing out the term *sport hunting*.

Safeguard hunting yourself. Hunters themselves can take the lead in the war against poaching. Don't wait for the wardens. First, armed with statistics about who the legal hunters are, tell your fellow hunters what the criminals are doing. Then tell your friends. Then tell schools, offices, churches, and communities. It will diffuse the antihunting outcry and help restore the positive image of the hunter. The hunter archetype will endure, as long as there are humans. The ethical hunter has two primary responsibilities: to act in accordance with Aldo Leopold's land ethic, and to help restore the hunter as a heroic archetype. If ethical hunters decline, poaching will increase.

Increase opportunities for people to hunt. Hunting is declining in many states because of complicated regulations, shorter seasons, expense, and a lack of land. Today, people don't hunt merely for sustenance. They hunt in order to have memorable experiences that will stay with them long after the year's last venison steak has been barbecued. Hunting preserves, in which hunters pay for what they shoot, will probably become more and more popular. On public lands, the growing numbers of hunters using bows and arrows and muzzleloaders need more support, because these people are the new ethical hunters. They don't take many animals; therefore, they should be given longer seasons in which to enjoy their sport. Kids, too, need special seasons and opportunities to hunt. In the years ahead, certain lands will need to be managed for primitive weapons, especially new urban areas where populations of deer, geese, and other species are growing and predators are scarce. Hunting with weapons that have a limited range is necessary in these zones. New hunting sanctuaries need to be created that can keep the passions of the hunt alive closer to home.

Notes

Introduction

1. This profile is based on statistics gathered from a 1991 U.S. Fish and Wildlife Service survey and a National Shooting Sports Foundation survey reported in the Washington Department of Wildlife hunter education program booklet "The Hunter in Modern American Society."

2. Sources: Safari Club International, U.S. Fish and Wildlife Service, Wildlife Management Institute, Whitetails Unlimited, Ducks Unlimited, and the National Wild Turkey Federation.

3. Daniel J. Decker, Jody W. Enck, and Tommy L. Brown, "The Future of Hunting—Will We Pass On The Heritage?" (paper presented at the second annual Governor's Symposium on North American Hunting Heritage, Pierre, SD, 24–26 August 1993).

4. J. Cravioto, et al., "Nutrition, Growth and Neurointegrative Development: An Experimental and Ecology Study," *Pediatrics* 38, pt. 2 (2) 1986.

5. Betsy Lehman, "Vegetarians on the Rise," *Marin Independent Journal*, 13 September 1993, sec. D, p. 1.

6. Marie-Louise von Franz, *Problems of the Feminine in Fairy Tales* (New York: Spring Publications, 1972), p. 56.

7. Stephen Kellert, "Attitudes and Characteristics of Hunters and Anti-Hunters" (transactions of the Forty-third North American Wildlife and Natural Resources Conference, 1978).

8. James Earl Jones and Penelope Niven, *James Earl Jones: Voices and Silences* (New York: Charles Scribners Sons, 1993), p. 23.

9. Bill Gilbert, "Hunting is a Dirty Business," *The Saturday Evening Post*, 21 October 1967.

10. Sam Keen, *Faces of the Enemy* (San Francisco: Harper and Row, 1986).

Chapter 1

1. Joseph Campbell, *The Way of Animal Powers* (San Francisco: Harper and Row, 1983), p. 73.

2. Dudley Young, *Origins of the Sacred: The Ecstasies of Love and War* (New York: St. Martins Press, 1991), p. 139.

3. Roger Williams, "Wolf Coursing" in *Hunting In Many Lands*, edited by Theodore Roosevelt and George Bird Grinnell. (New York: Forest and Stream Publications, Inc., 1895), p. 327.

4. William James, *The Varieties of Religious Experience* (New York: Random House, 1902), p. 66.

5. Ted Nugent, "A Double On Big Kahunas" in *Ted Nugent World Bowhunters*, December 1992/January 1993, p. 20.

6. Michael Murphy, *The Future of the Body: Explorations into the Further Evolution of Human Nature* (Los Angeles: Jeremy Tarcher, 1992).

7. Mircea Eliade, *The Sacred and the Profane* (New York: Harcourt, Brace and Jovanovich, 1959), p. 11.

8. Charles Joy, ed., *The Animal World of Albert Schweitzer: Jungle Insights Into Reverence for Life* (Boston: Beacon Press, 1950), p. 165.

9. Erich Fromm, *The Anatomy of Human Destructiveness* (New York: Holt, Rinehart and Winston, 1973), p. 132.

Chapter 2

1. Carleton Coon, *The Hunting Peoples* (Boston: Atlantic-Little Brown, 1971), p. xvii.

2. José Ortega y Gasset, *Meditations on Hunting* (New York: Scribner's, 1972).

3. Erich Fromm, *The Anatomy of Human Destructiveness* (New York: Fawcett, 1975), p. 160.

4. S. Boyd Eaton, Marjorie Shostak, and Melvin Konner, *The Paleolithic Prescription: A Program of Diet and Exercise and a Design for Living* (New York: Harper and Row, 1988), p. 4.

5. Mircea Eliade, *Shamanism: Archaic Techniques of Ecstasy* (Princeton, NJ: Princeton Univ. Press, 1966), p. 87.

6. A. Foster, "ESP Tests with American Indian Children," *Journal of Parapsychology*, 7 (1943): pp. 94–103.

7. Robert Ardrey, *The Hunting Hypothesis* (New York: Athenum, 1976).

8. Aniela Jaffe, "Symbolism and the Visual Arts" in *Man and His Symbols*, ed. Carl Jung (New York: Dell, 1968), p. 266.

Chapter 3

1. Thomas Merton, ed., *Gandhi on Nonviolence* (New York: New Directions, 1964), p. 11–69.

2. This profile is drawn from data collected by Dr. Jon Hooper, reported in "Animal Welfarists and Rightists: Insights into an Expanding Constituency for Wildlife Interpreters," in *Legacy*, November/December 1992, pp. 20–25.

NOTES   278

3. This statement, and others which follow from different organizations, are excerpted from "What They Say About Hunting: Position Statements on Hunting of Major Conservation or Preservation Organizations," National Shooting Sports Foundation, 555 Danbury Road, Wilton, Connecticut 06897.

4. Kathleen Marquardt, *Animal Scam: The Beastly Abuse of Human Rights* (New York: Regnery, 1993), p. 28.

5. George LaPointe, "Proactive Strategies Project Executive Summary of Regional Workshops" (Baton Rouge, LA: Louisiana Department of Wildlife and Fisheries, 1991).

6. Jon Hooper, *Animal Welfarists and Rightists: Insights into Expanding Constituencies for Wildlife Managers* (U.S. Department of Defense, 29 September 1993).

7. Sigmund Freud, *Group Psychology and the Analysis of the Ego* (New York: Norton, 1959), p. 20.

8. Aldo Leopold, *A Sand County Almanac* (New York: Oxford Univ. Press, 1966), pp. 219–20.

9. Sam Keen, *Faces of the Enemy* (San Francisco: Harper and Row, 1986), p. 19.

10. Duncan Barnes, "Up Front," *Field and Stream*, March 1993, p. 5.

11. Rod Strand and Patti Strand, *The Hijacking of the Humane Movement* (Wilsonville, OR: Doral Publishing, 1993).

12. Richard Nelson, *Make Prayers to Raven: A Koyukon View of the Northern Forest* (Chicago: Univ. of Chicago Press, 1983).

13. Michael Crichton, *Congo* (New York: Ballantine Books, 1980), p. 150.

Chapter 4

1. Erik Erikson, *The Life Cycle Completed* (New York: Norton, 1982).

2. Peter Singer, *Animal Liberation*, 2nd ed. (New York: Random House, 1990), p. 159.

3. Marie-Louise von Franz, *Problems of the Feminine in Fairy Tales* (New York: Spring Publications, 1972), p. 188.

4. von Franz, *Problems of the Feminine in Fairy Tales*, p. 185.

5. George Schaller, *The Serengeti Lion*, 1972.

6. Erich Fromm, *The Anatomy of Human Destructiveness* (New York: Holt, Rinehart and Winston, 1973), p. 125.

7. S. L. Washburn, in Robert Ardrey's *The Hunting Hypothesis*, p. 21. Washburn has written extensively about the anthropology of hunting. See S. L. Washburn and C. S. Lancaster, "The Evolution of Hunting" in *Man, The Hunter*, ed. R. B. Lee and I. DeVore (Chicago: Aldine, 1968); and S. L. Washburn, ed., *The Social Life of Early Man* (Chicago: Aldine, 1961).

8. Stephen King, *The Wastelands* (New York: Signet, 1993), p. 101.

Chapter 5

1. S. L. Washburn and C. S. Lancaster, "The Evolution of Hunting" in *Man, The Hunter*, ed. R. B. Lee and I. DeVore. (Chicago: Aldine, 1968).

2. Rudolph Otto, *The Idea of the Holy* (New York: Oxford Univ. Press, 1982), pp. 14–15.

3. H.R. 1815, 103rd Congr., protects "individuals engaged in a lawful hunt on Federal lands, to establish an administrative civil penalty for persons who intentionally obstruct, impede, or interfere with the conduct of a lawful hunt." Monies collected from such cases are to be used for habitat protection and enhancement.

4. Wildlife Management Institute, "Placing Hunting in Perspective" (Washington, DC GPO, 1992), p. 22.

5. This August 1990 public opinion study was commissioned by the National Shooting Sports Foundation and conducted by the Gallup Organization, Inc.

6. Stephen R. Kellert, "Attiudes and Characteristics of Hunters and Antihunters" (transactions of the 43rd North American Wildlife and Natural Resources Conference, 1978), pp. 412–23.

7. Lewis Regenstein, *The Politics of Extinction: The Shocking Story of the World's Endangered Wildlife* (New York: Macmillan, 1975), p. 39.

8. Karl Menninger, "Totemic Aspects of Contemporary Attitudes Toward Animals," in *Psychoanalysis and Culture*, ed. G. B. Wilbur and W. Muensterberger (International Universities Press, 1951), p. 45.

9. Menninger, "Totemic Aspects," p. 45.

10. Lucy Freeman, ed., *Karl Menniger, M.D.: Sparks* (New York: Thomas Y. Crowell, 1993), p. 246.

11. Erich Fromm, *The Anatomy of Human Destructiveness*, pp. 131–32.

12. Karl Menninger, *A Psychiatrist's World*, ed. B. H. Hall (New York: Viking Press, 1959), pp. 77–78.

13. Karl Menninger, *The Crime of Punishment* (New York: Viking Press, 1966), p. 171.

14. Karl Menninger, *The Vital Balance* (New York: Viking Press, 1963).

15. Joseph Wood Krutch, quoted in Holmes Ralston III, *Environmental Ethics: Duties to and Values in the Natural World* (Philadelphia: Temple Univ. Press, 1988), p. 78.

16. Carleton Coon, *The Hunting Peoples* (Boston: Atlantic-Little, Brown, 1971), p. 3.

17. Bill Karr, "Deer, Cats, Trophies and More on Animal-Rightists," *Western Outdoor News*, December 1993, p. 2.

18. Yuri Simchenko, "Nganasan Shamanism: Snow and Reindeer," in

Shamanism (Norwalk, CT: Foundation for Shamanic Studies, 1993) spring/summer 1993, vol. 5 no. 4 and vol. 6 no. 1, pp. 17–22.

19. Quoted in *The New Hunter's Encyclopedia* (Harrisburg, PA: Stackpole Books, 1966).

20. Aldo Leopold, *A Sand County Almanac* (New York: Oxford Univ. Press, 1966), p. 240.

21. California Department of Fish and Game, *California Hunter Education Manual*, rev. ed., (Sacramento, CA: 1987).

22. Richard Nelson, "The Gifts," in *On Nature: Nature, Landscape and Natural History*, ed. Daniel Halpern (San Francisco: North Point Press, 1987), pp. 117–31, 124.

Chapter 6

1. "Clinton Hunts, Making a Point on Guns," *New York Times*, 27 December 1993, sec. A. p. 7.

2. Henry David Thoreau, *Walden*, in *The Portable Thoreau*, ed. Carl Bode (New York: Viking Press, 1947), p. 459.

3. Robert Graves, *The White Goddess* (New York: Farrar, Strauss, Giroux and Cudahy, 1948), pp. 16–21.

4. Mircea Eliade, *The Sacred and the Profane* (Princeton, NJ: Princeton Univ. Press, 1964).

5. Andreas Neher, "A Physiological Explanation of Unusual Behavior in Ceremonies Involving Drums," *Human Biology* 34 (1962): pp. 51–160.

6. Eliade, *The Sacred and the Profane*.

7. State of California Department of Fish and Game, *Deer Hunting* (draft environmental document), 12 February 1993, p. 188.

8. Awo Fa'lokun Fatunmbi, *Ogun, Ifa and the Spirit of Iron* (Bronx, NY: Original Publications, 1992).

9. Browne and Williams, "Resource Availability for Women at Risk: Its Relationship to Roles of Female Perpetuated Partner Homicide" (paper presented at the annual conference of the American Society of Criminology, 1987).

10. Jimmy Carter, *An Outdoor Journal: Adventures and Reflections* (New York: Bantam, 1988), p. 39.

11. California's crime statistics taken from several *Marin Independent Journal* articles, January 1993.

12. James Swan, *Nature as Teacher and Healer* (New York: Villard-Random House, 1992), pp. 173–75

13. D. T. Suzuki, foreword to *Zen in the Art of Archery*, by Eugen Herrigel (New York: McGraw-Hill, 1964).

Chapter 7

1. Marie-Louise von Franz, *Problems of the Feminine in Fairy Tales* (New York: Spring Publications, 1972), p. 34.

2. James A. Swan, *Nature as Teacher and Healer* (New York: Villard-Random House, 1992).

3. Merritt Clifton, "Killing the Female: The Psychology of the Hunt," *Wingspan*, fall 1991, p. 9.

4. Matt Cartmill, *A View to Death in the Morning: Hunting and Nature Through History* (Cambridge, MA: Harvard Univ. Press, 1993). Provides an in-depth, scholarly examination of the mythology of hunting.

5. Margaret L. Knox, "In the Heat of the Hunt," *Sierra*, November/December 1990, pp. 48–59.

6. Melvin Konner, *Why the Reckless Survive: And other Secrets of Human Nature* (New York: Viking Press, 1990), p. 4.

7. Christine L. Thomas and Tammy A. Peterson, "Becoming An Outdoors-Woman: Concept and Marketing," 1993.

8. The Tomboy Club, P.O. Box 846, Dallas, OR 97338. For a women hunters' clothing catalog, contact: Lady Hunter, Dingman's Ferry, PA 18328, 1-800-241-4845.

9. Tim Norris, "Women Hunters Show They Aren't A One Shot Deal," *Milwaukee Journal*, 7 November 1993.

10. Carol J. Adams, *The Sexual Politics of Meat: A Feminist Critical Theory* (New York: Continuum, 1990).

11. David Carradine, *The Spirit of Shaolin* (Boston: Charles E. Tuttle Company, 1991), p. 139.

12. Marie-Louise von Franz, *Individuation in Fairy Tales* (Boston: Shambala Publications, 1990), p. 12.

13. Congressional Sportsmen's Congress, "Release of Captive Mallards—Let's Take a Good Look," Blanch Lambert (Washington, D.C., 1989).

Chapter 8

1. Henry David Thoreau, "Higher Laws" from *Walden* in *The Portable Thoreau*, ed. Carl Bode (New York: Viking Press, 1947), p. 458.

2. Bill Gilbert, "Hunting Is a Dirty Business," *The Saturday Evening Post*, 21 October 1967.

3. Joseph Campbell, *The Masks of God: Primitive Mythology* (New York: Viking/Penguin, 1959), p. 291.

Index

Adams, Carol J., 243
Addiction, 28, 224
Adolescence: hunting in, 140–41; initiation rites in, 135–36. *See also* Juveniles
Aggression, 215, 218; Fromm on, 65–66, 136, 215
Ahimsa doctrine, 122
Ainu, bear sacrifice by, 81–84
Akuette, Durbach, 84
American Humane Association, 93
American Humane Society, 162
Amory, Cleveland, 104, 108–9, 160, 161
The Anatomy of Human Destructiveness (Fromm), 66, 169
Animal Liberation (Singer), 116, 131
Animal Liberation Front (ALF), 93–94
Animal rights activists, 9–10; and animals, 116; antihunting activities of, 10–11, 22, 156–57, 158, 163; and fish and game managers, 107–8; mental health of, 114–15; motivations of, 188; public opinion of, 163
Animal rights movement, 110–11, 113–14, 218; average supporter of, 93; conference of, 99–104, 107, 108–9, 110; and ethics, 111–13, 119; growth of, 115–16; heroes of, 121–22; organizations in, 93–94, 99; questions for, 118–21; responsibility of, 117
Animals: as conduit to sacred, 35; dreams of, 45–47, 48; extinct, 5–6, 53–54; human encounters with wild, 97–98; and hunters' success, 37–38; increasing populations of, 6–7, 159–60; killed yearly by hunters, 7–8; as messengers and guides, 32, 36–37, 41, 90, 91; politically correct actions with, 99; road-killed, 89–90; sacrifices of, 82–84; traditional cultures' relationship with, 59–62, 121
Antihunters, 10–11, 87, 171–72, 271–72; accusations of, 158, 160, 162, 163, 164, 166, 173, 175; motivations of, 171–72
Archery. *See* Bows and arrows; Bow hunting
Ardrey, Robert, 71
Arkansas, 163
Armstrong, Thomas, 223
Art: decoys as, 51–52; totem poles as, 59–60; wildlife, 144
Artemis, 179
Assault weapons, 220
Athletes, 27–28, 240–41
Atlatls, 200
Auel, Jean, 241
Azatlan, Indian mounds at, 32

150–51; oil-coated, 193–94; population level of, 164–66; and wetlands loss, 7; wood, 6

Ducks Unlimited, 8, 22, 115, 159, 165

Eastwood, Clint, 64, 170, 197

Eaton, S. Boyd, 66

Economy, hunter contribution to, 73. *See also* Cost

Eliade, Mircea, 35, 69, 203

Elk, 6, 8; tule, 88–89, 91–92, 94, 95–97

Emotions: and hunters, 238–39, 240; and psychology, 228–29. *See also* Mental health

Enck, Jody, 12

Erikson, Erik, 129, 131, 140

Ethics: and animal rights movement, 111–13, 119; Fair Chase, 181; hunting, 188–89; land, 112, 182

Evil, 173–75

Extinction, 5–6, 53–54

Extrasensory perception, 69

Fair Chase ethic, 181

Fear, 35, 113, 116, 130, 142

Federal Aid in Wildlife Restoration Act, 159

Feminine, 245; Jung on, 228; potential pathology of, 233; repression of, and illness, 211, 236–38. *See also* Women

Feral pigs, hunting for, 262–66

Finson, Bruce, 37–38

Fish and game managers, 107–8

Fishing, 26, 126

Fishing licenses, women buying, 241

Food supply: modern, versus wild game, 66–67; today's, 13–14. *See also* Meat eating; Vegetarianism

Foster, A., 69

Franz, Marie-Louise von, 16, 46, 131–32, 226, 247

Freeman, Lucy, 168

Freud, Sigmund, 13, 110–11, 168, 174, 177, 197, 206, 234

Friends of Animals, 99, 162

Fromm, Erich, 49, 66, 136, 169, 170, 215

Fund for Animals, 102, 103, 104

Gallup Poll, 163

Gandhi, Mahatma, 9, 35, 88, 121

Gangs, 222–23

Geller, Uri, 227

Gilbert, Bill, 18, 255

Goodall, Jane, 139–40, 177

Government: hunting regulations of, 8, 180–81, 253–54, 258, 259–60, 266–67; revenues from hunting to, 73, 159

Graves, Robert, 201

Grinnell, George Bird, 28, 147

Grizzly Island National Wildlife Refuge, 94

Guilt, 29, 64–65, 238–39

Gunpowder, 208–9

Guns: in America, 113, 195–96; childhood experiences with, 190–95; hunting, 195–97; making of, 210–11; murders with, 214–15; symbolism of, 168–69, 197, 212–13; using shells, 209

Habitat, 75, 115

Handguns, 209

Hayden, Tom, 100

Healers, and killing, 23

"High," 34–35, 130, 141

Himeda, Tadayoshi, 82

Hooper, Jon, 108

Hughes, Robert, 217

Huichol Indians, 37–38, 52, 78, 184, 196, 203, 229

Humane Society of the United States, 93, 162–63

Humans: denial of inner nature by, 13, 15, 131–32, 176; evolution of,

life Refuges, hunting in, 257–60, 267

Sacred, deer as, 80–81. *See also* Spirituality

The Sacred and the Profane (Eliade), 35

Sadism, and hunting, 166, 167, 169–71, 172, 235, 236

Safari Club International, 8, 273

A Sand County Almanac (Leopold), 112, 182

Santeria, 84

Saper, Joel, 166

Scapegoats, hunters as, 12, 111, 116

Schaller, George, 134

Schindler's List, 170

Schools, 223

Schweitzer, Albert, 9, 44, 122, 172

Sea Shepherd Society, 103

Self-actualization, 35, 127, 171

Self-Esteem Task Force, 215

Senses, in traditional cultures, 67–70

Sequoia, Anna, 160

The Sexual Politics of Meat (Adams), 243

Sexual symbolism: of guns, 168–69, 197; in hunting, 231, 234–36

Shamans, 37, 121, 164; and bows and arrows, 203

Shostak, Marjorie, 66

Sierra Club, 115, 147

Singer, Peter, 110, 111, 116, 131, 132, 163

Slings, 207

Slob hunters, 18, 168, 247

Slugfest (Guerneville, Calif.), 9–10

Snow geese, 39–41, 42–43; dreams about, 45–46, 47

Sparks (Freeman), 168

Spears, 199–200

The Spirit of Shaolin (Carradine), 245

Spirituality, 79; in eating game, 23; in hunting, 35, 155–56, 239; in nature, 57–58; peak experiences in, 34–35; of weapons, 196–97

Sport hunting, 144, 162–63, 274; in media, 255

Sports, peak experience in, 34

Sportsmen Against Hunger, 273

Stalking, 68; Ortega y Gasset on, 63–64

Stevens, Christine, 10–11

Subsistence hunting, 17, 127; in Alaska, 85–86. *See also* Traditional hunters

Sundberg, Norman, 105

Suzuki, D. T., 225

Swan, James, 124

Tenrikyo religion, 60–61

Thomas, Christine, 242

Thoreau, Henry David, 122, 190, 249

Tieder, Kathy, 246

Tiller, William, 227

Time, and hunting, 31–32

Tomboy, 242

Totem, 46

Totem poles, 59–60

Tracking, 68

Traditional cultures: animal's choice in, 37–38; bear in, 81–84; deer in, 77–79; diet of, 66–67; intuition in, 229; kinship of, with animals, 59–62; mental health of, 229; paranormal in, 68–70

Traditional hunters: guides for, 178–80; nature as spiritual for, 57–58; preparation by, 58–59, 63; senses of, 67–69; weapons and techniques of, 65–66, 67, 181–82. *See also* Subsistence hunting

Trees, old-growth, 57–58